RESEARCH THAT S

THE RESEARCH OPERATIONS HANDBOOK

Kate Towsey

 Rosenfeld®

NEW YORK   2024

Research That Scales
The Research Operations Handbook
By Kate Towsey

Rosenfeld Media, LLC

125 Maiden Lane

New York, New York 10038

USA

On the Web: www.rosenfeldmedia.com

Please send errata to: errata@rosenfeldmedia.com

Publisher: Louis Rosenfeld

Managing Editor: Marta Justak

Interior Layout: Danielle Foster

Cover Design: Heads of State

Illustrator: Danielle Foster and Nataliia Kokhanchyk

Indexer: Marilyn Augst

Proofreader: Sue Boshers

To my husband—
the countless hours you've spent listening,
reading, and brainstorming with me mean that you are,
by proxy, an expert in ResearchOps.
You are present on every page.

HOW TO USE THIS BOOK

Who Should Read This Book?

Research That Scales is a resource for anyone involved in devising the strategies and systems that support human-centered research. More than just tips and checklists, this book will stretch your sense of what it means to run a research practice and give you practical frameworks and examples for how to *think* operationally so that you can make the right moves.

ResearchOps requires that craft leaders and operational allies are a tightly coordinated duo. So, whether you're a leader, a researcher, or an ops specialist working in the thick of things, this book will help you level set. Use it to get everyone on the same page (literally) about the nature of the work and what to do next.

A wide variety of disciplines do research these days and also need robust and scalable research operating systems. If you work in product, design, content, marketing, academia, or for an agency, and you do research as part of your work, these pages hold equal value for you. So that everyone feels welcome, I've intentionally used terms like *research practice* rather than *research team*, where appropriate.

Finally, the economic shocks of the post-pandemic era mean that resiliency is top of mind. In places, this book responds to the context in which it was written (2020–24), but the concepts are timeless. Whether you're leading a team that runs into the hundreds or you're just starting out, the advice shared in this book will help you build a research practice that can thrive when times are good and weather inevitable storms.

What's in This Book?

Research That Scales is a complete guide for approaching research operations both strategically (top-down) and tactically (bottom-up). In keeping with this pattern, it is structured in two parts.

Chapters 1–4 demystify the key terms *scaling research* and *research strategy,* and cover fundamental concepts that underpin research operations. By the close of these chapters, you'll have a pragmatic understanding of what it means to scale research, how to build purpose into your practice, and how to plan operations that make good ideas real. These chapters provide a crucial backdrop for the specialized work explored in the following eight chapters, so you should read them first.

Chapters 5–12 are dedicated to the *eight elements of ResearchOps*, one for each element. You'll be introduced to the full list of elements in Chapter 4, "Planning Realistic ResearchOps." Each "elemental chapter" is a deep dive into an element, say participant recruitment or ethics and privacy, and offers a wealth of frameworks, stories, and case studies so that concepts are clear and highly applicable, whatever your context.

I've written this book with a sequence in mind, but you needn't read it from start to end. Ideally, read Chapters 1 to 4 first and in sequence, and then you might dip in and out of the eight "elemental" chapters, i.e., Chapters 5 through 12, as needed.

What Comes with This Book?

This book's companion website (☍https://rosenfeldmedia.com/books/research-at-scale) contains a blog and additional content. The book's diagrams and other illustrations are available under a Creative Commons license (when possible) for you to download and include in your own presentations. You can find these on Flickr at www.flickr.com/photos/rosenfeldmedia/sets/.

FREQUENTLY ASKED QUESTIONS

Isn't ResearchOps just about procuring the right tools and having someone take care of participant recruitment for me? Is an entire book necessary?

There's no doubt that procuring tools and full-service participant recruitment are key research operations themes, but it's a mistake to think that these tasks constitute ResearchOps. Operations is often confused with administration or assistance, but administering systems or assisting researchers is just one part of the job. As you will learn in **Chapter 1**, "Research Does *Not* Scale—Systems Do," to help an organization operate in ways that add value, you'll need to deliver the right systems first and then administer them well. If you deliver administrative support before designing systems, you're putting the cart before the horse.

ResearchOps sounds like just another buzzword, like DevOps and DesignOps. What does "Ops" even mean, and where does it come from?

Operations, or simply *Ops*, is a buzzword that's been at the forefront of software development and experience design for years. *DevOps* emerged sometime around 2007. *DesignOps* in 2015. And, in 2018, *ResearchOps* leapt into the awareness of researchers globally. Relative to their operational forebears, DevOps, DesignOps, and ResearchOps are young specialisms that owe much to a long line of pioneers: the industrialists, analysts, organizers, and strategists who began trailblazing the world of operations during the first half of the twentieth century. Operations has a long and diverse history, but, whatever the field of interest, the goal has always been the same: to enable "the power to act."[1] To foster excellence in execution. To maximize efficiency, profit, and value where it's needed most. And to support the ability to scale.

---

1   Marco Iansiti, "The History and Future of Operations," hbr.org, June 30, 2015, https://hbr.org/2015/06/the-history-and-future-of-operations

AI has taken the world by storm, and it's a hot topic in research. Why haven't you written a chapter about it?

It goes without saying that artificial intelligence (AI) has impacted every part of society, and research is no exception. Large language models (LLMs) like ChatGPT and Google's Bard (now called Gemini) are powerful tools with the potential to super-charge a research practice if used well, but LLMs are also just tools. As covered in Chapter 8, "Tactical Tooling," *every* tool that you onboard should serve a particular purpose and jibe with your systems and people, and AI is no different.

> **NOTE**
> ChatGPT agrees with this statement—I checked!

I can see the value of building a dedicated ResearchOps team, but I need to get others onside. Will this book help me shape a business case?

Yes, yes, and *yes*. One of the biggest mistakes that folks make when they're pitching ResearchOps for the first time, is not defining a strategy for what they want to achieve and how they will achieve it. Chapters 1–4 will help you understand how to define a research strategy and how to plan operations that deliver measurable value to the business—things that purse-string holders care about.

When it comes to delivering something specific, like a research library or a participant recruitment panel, Chapters 5–12, i.e., the "elemental" chapters, will help you make your business case specific. For example, Chapter 5, "People to Take Part in Research," covers the nitty gritty of scaling participant recruitment. Chapter 6, "Long Live Research Knowledge," is a deep dive on research knowledge management, like libraries and repositories. And Chapter 10, "Money and Metrics," will help you understand how to use metrics to bolster your case.

Will investing in ResearchOps help future-proof my research team?

When the going gets tough, even highly valued people and partners are sometimes let go. Even so, the more a research practice can deliver reliable and tangible value to the business, the more resilient it will be. Crucial to delivering value is devising a research strategy that addresses the things executives care about and that builds core strength—like Pilates for research—within the team. All the content in this book will help you to future-proof your team because building resilience (not just size) is what scaling research is all about.

CONTENTS AT A GLANCE

CONTENTS

FOREWORD

Having built successful multidisciplinary technology innovation and research teams in many corporations and academic contexts, I'm aware that having a clear research operations approach and strategy is critical. Here's one of my favorite examples:

When Google's highly successful Material Design (now called *Google Design Platform*) was launched in 2014, there was no UX research on the design guidelines, on the designed components, or on the impact of Material Design as a whole. In collaboration with two colleagues, I proposed building a research team for Material Design and scoped out an initial program of work—a big shout-out to Dave Chiu, Staff Interaction Designer, and Michael Gilbert, Staff UX Researcher, both still at Google, who were on that journey with me. After a few key successes, I was given the head count to expand the team and hired qualitative and quantitative researchers and data engineers.

Using the kinds of approaches that Kate outlines in *Research That Scales*, and with the help of key hire, program manager Zaina Alhawi, now Senior UX Programs & Operations at Google, a highly impactful research team was established. We scaled from 3 to 18 team members with programs of work deeply integrated with the Material Design PM and engineering functions. Further, through our newsletter, consulting, and communications programs we had influence and impact well beyond our own Material Design context, advising on research programs and design changes in product teams within and outside of Google. Years later, the research team remains a central part of Google's Design Platform. As Kate points out, *scale* is a rich word with many facets.

Key to the success of the entire team and the value proposition for Google was having a clear research operations strategy focused on problem framing, solid research insights, and a deep commitment to reinforcing the leadership collaboration between research, design, engineering, and product management. Using many of the techniques that Kate suggests in *Research That Scales*, we created a systematic, "operations-oriented" research program.

If you're creating or running research for any company, small or large, you too need to have a set of operating principles that defines the problem space, identifies information gaps and opportunities, and offers an approach for igniting interest in others—especially stakeholders, partners, and collaborators.

So, I am excited for you to read *Research That Scales*. Brava to Kate for bringing her experience and insights together in this book! Whether you're new to the world of research operations or an experienced leader, there are provocations, models, metaphors, and frameworks here for you to take on board.

Happily, the acknowledgment of the importance of research operations has grown over the last five years, in no small part owing to Kate's efforts. Years ago, Kate saw the need to bring together people struggling with these issues; she popularized the term *ResearchOps* to describe what we were all doing, but often doing alone. First, through her community work, and now with this book, Kate has offered a host of people the space to come together to share inspirations and pain points, and to become more effective leaders, pioneers, and innovators in sustainable UX research practices.

—Elizabeth F. Churchill, PhD
Department Chair and Professor, HCI/UX & AI, MBZUAI, UAE
Former Google Senior Director of UX

INTRODUCTION

How to Eat an Elephant

In March 2018, I posted a tweet that said "I've started a ResearchOps Slack channel[2] so we've got a place to share, learn, and be geeky about #userresearch operational stuff. Let me know if you want to join: #ResearchOps."[3] The tweet had 12 reposts and 56 likes—limited engagement by ordinary standards—but it set the wheels in motion for ResearchOps to become a bona fide global profession. (Oh, the good old days of Twitter!) It's mind-boggling to think that, in just six years and despite a pandemic and an economic shakedown of note, hundreds of people around the world are now employed as ResearchOps professionals. No longer a novel function for the elite, ResearchOps is an established part of the research vernacular and a dedicated role in just about every scaled-up or scaling research team.

While I'm regularly credited with inventing research operations, it's not true. I did put ResearchOps on the map by giving it a community,[4] a voice, a definition, a framework,[5] and then a professional members' club[6] (and now this book). But formal ResearchOps teams have existed for decades, albeit within the largest and most forward-thinking companies of our time. Microsoft, for instance, has had a ResearchOps function in one form or another for 20 years, and tech behemoths Google, Facebook, Salesforce, and Airbnb have had ResearchOps in their midst for at least a decade. Two of the blurb writers for this book, Tim Toy and Noel Lamb, have worked in research operations roles since 2008-9. Sure, people with this length of experience are few and far between, but they are the pioneers—the living legends.

---

2 The Slack channel was the ResearchOps Community, which I stepped away from in 2019.

3 X (formerly Twitter), https://x.com/katetowsey/status/971754907974938624

4 The ResearchOps Community, https://researchops.community/

5 A framework for #WhatisResearchOps by Kate Towsey: https://medium.com/researchops-community/a-framework-for-whatisresearchops-e862315ab70d

6 Since 2019, I have run a members' club for ResearchOps professionals. Unconventionally, it's called the Cha Cha Club. https://chacha.club/

So, if ResearchOps was barely known in 2018, where have all the legions of people who now work as ResearchOps professionals come from?

The beauty of ResearchOps is that it doesn't require new skills. Instead, it requires a recombination (or reimagining) of all sorts of existing systems—and people-oriented know-how—plus bucket loads of entrepreneurial wisdom and grit! People come to ResearchOps from all sorts of backgrounds, the most common are social sciences, hospitality, human resources, business management, marketing, and, of course, research. But service and systems design and the information sciences are also invaluable allies, and I hope more folks from those disciplines will join the fold—and that you will hire them.

It might come as a surprise that I came to operations with a background in Fine Art. My degree centered on conceptual art, the goal of which isn't to visually impress—sometimes these works are quite ugly—but rather to incite the viewer to comprehend or feel something new: to offer *insight*. To achieve this, an artist needs to work out a conceptual system or framework for each art work or exhibition. In other words, they must understand the images, shapes, materials, textures, colors, sounds, smells, or words—the complete experience— that will evoke a particular response. I wouldn't outright recommend a Fine Art degree as a precursor to ops,[7] it's a bit left field, but the simultaneously creative, systematic, and logical thinking it requires is a surprisingly good fit when your work centers on building cognitive systems—or *learning systems*.

If this book (and that story) achieves anything, I hope it empowers you to draw a diversity of powerful, if not obvious, skills into your research practice, and then use them well. To scale research, you'll need to know when and how to spot, coordinate, and apply specialist skills, and for what purpose. I wrote this book to fill *that* knowledge gap and to stretch your sense of what research and research operations is all about.

---

7   Information science, librarianship, archiving, an MBA, or HCI are all good ops degrees. Service design, human-centered design, and user research are also useful fields to study.

Operations is as much an art as it is a science, which is what makes the work so endlessly interesting. "It depends" is an ever-present theme, so you'll need to be equal parts creative and pragmatic at every turn. So, although I've included several models, frameworks, and lists, none of them are prescriptive or rigid. Instead, the intent is to help you *think* operationally so that you can deliver research systems that are custom-fit and responsive, whatever your context.

Finally, it's worth noting that the stories I share are often big or scaled up. It *is* a book about scaling research, after all! Everything that I share, I've delivered in real life to dozens or hundreds of people, over several years and with lots of trial and error on the way—errors and inefficiencies you'll be able to avoid, thanks to this book. If you're working at small scales or you're just starting out, don't be discouraged by the enormity of the task. Instead, see the contents of this book as a vision of where you *can* get to and a map for *how* to get there in manageable steps. As Desmond Tutu, the late South African archbishop and Nobel Peace Prize laureate said, "There is only one way to eat an elephant: a bite at a time."

CHAPTER 1

Research Does *Not* Scale— Systems Do

It was a sunny day in 2018, and from an upper floor of the Goldman Sachs building, I could see the Statue of Liberty standing strong across the Hudson River—a quintessential New York City moment. I took a deep breath and gathered my thoughts, then entered a room full of fifty research leaders who had signed up to take part in a one-day workshop that I was facilitating for the first time; the workshop was called *Designing a ResearchOps Strategy*. To prepare for the day, I had asked attendees to come armed with notes about their research strategy because it would be the chief informant of their strategy for ResearchOps. On that day, though, I discovered that only a handful of the research leaders who had signed up to take part actually understood what a research strategy was or had one in place. This wasn't a one-off.

Of the ~300 leaders who have attended my workshops over the past five years, few have had an established research strategy or felt comfortable that they knew what one was, yet they all wanted to scale research. To scale anything, you must know with precision what you are going to scale, and why. To sustain anything—scalable things are *always* sustainable—you must know when to invest, what the long-term investment is, and when to let sleeping dogs lie. To coordinate complex operations—scalability can't be delivered in silos; it requires a systems-thinking approach—you will need a game plan for what you want to achieve and how you'll succeed. In other words, you'll need a synchronous research and research operations strategy, and the know-how to deliver them successfully.

This is an introduction that pulls no punches, and it is littered with words like *scale*, *operations*, *systems*, and *strategy*. So, before getting lost in the minutia of scaling participant recruitment, knowledge management, or research ethics, to name just a few of the jobs of research operations, it is essential to exchange the jargon, trends, and clichés, which have for the most part been inherited from the world of manufacturing, for a pragmatic and comprehensive understanding of what these business words mean—*really* mean—both broadly and in the context of research.

> **NOTE** FROM CASUAL CHATS TO PEER-REVIEWED PAPERS
>
> From ad hoc chats with customers to published papers that are peer-reviewed, the word *research* is used to describe a broad spectrum of activities, assets, and levels of rigor: its purview is the fuel of endless industry debate. In this book, I've used the

word *research* to describe a broad spectrum of knowledge-gathering, or learning, activities. The levels of rigor that are suited to your context are up to you or your leaders to decide, and it's this sort of decision making that forms the basis of a good research strategy. (See Chapter 2, "Lost and Won on Strategy.")

Demystifying *Scale*— *The* Business Buzzword

Within the last week, you've likely taken part in or overheard a conversation about scaling something up or down, making something scalable, or delivering something to (or at) scale. It is a business cliché that's strewn across websites, conversations, documents, and presentations. For instance, a popular research technology company promises that its platform can help you "scale customer insights with AI-powered research." Putting the words "AI-powered research" aside—they're important, too, but that's a discussion for another place and time—what does "scaling customer insights" mean? They are words that are often said and printed in marketing blurbs, but where would you start? What value would you seek to achieve? How would you measure it? These are important questions to ask as well. Answering them requires momentarily putting the word *research* aside to unpack the meaning of the word *scale*, not just theoretically but pragmatically.

A Dictionary Definition

If you were to consult a dictionary, you would find that *scale* has countless relevant meanings, such as a machine for weighing, a series of marks for measurement, to scale (or climb) a mountain or ladder, for instance, or to grow in extent or indicate extent—the damage is large-scale. But the business world has evolved the word beyond its original meaning. In a blog post, the Merriam-Webster dictionary wrote: "In the past few decades, *scale up* ('to increase the size, amount, or extent of something') has been shortened to *scale* when used in business contexts…. This newer use of scale means 'to grow or expand in a proportional and usually profitable way.'"[1]

---

1 "What Does 'Scale the Business' Mean? How a Common Word Became a Staple of Business Jargon," Merriam-Webster, www.merriam-webster.com/words-at-play/scale-the-business-meaning-origin

It's a good-as-any dictionary definition, but dictionaries are built for brevity and not applicability, so a more streetwise definition is needed.

Scale Isn't About Large Numbers

The best viewpoint about what *scale* means comes from the technologist Archis Gore. In his blog post, "You're Thinking About Scale All Wrong,"[2] a pointed title if there ever was one, he says: "A scalable system isn't one that launches some fancy large number and just stupidly sits there. A scalable system is one that scales as a verb...."

To put that in the context of research, imagine that you've delivered a system that can send thank-you gifts, say egift cards or swag, to 100 research participants per week. For the system to be scalable, it should be able to reliably deliver the same level of service as demand rachets up, say from 100 to 400 thank-you gifts per week. This doesn't mean that a scalable system should handle increasing levels of demand without further investment. Rather, to scale a system, you must understand the proportional adjustments that you'll need to make in your supply—money, software seats, staff, participants, and that most crucial resource, *time*—to reliably deliver the same level of service to increasing numbers of people (see Figure 1.1).

"Calling something 'scalable' simply because it is very, very, very large is like calling something realtime [sic] only because it is really, really fast...." Gore beautifully drives the point home. He adds, "Did you know that nowhere in the definition of 'real-time systems' does it say 'really, really fast?' Real-time systems are meant to be time-deterministic, i.e., they perform some operation in a predictable amount of time." Indeed, the most common words that crop up to define scale, whether in the business context or otherwise, are "predictable," "proportional," and "measured," rather than "large."

So, scale has little to do with size and everything to do with known proportions, predictability, and control. That is the first point to get right, but there is another business buzzword that drives misassumptions about scaling things: *efficiency*.

---

2   Archis Gore, "You're Thinking About Scale All Wrong," *Archis's Blog: Ramblings of a Perpetually Bored Person* (blog), June 5, 2018, https://archisgore.wordpress.com/2018/06/05/youre-thinking-about-scale-all-wrong/

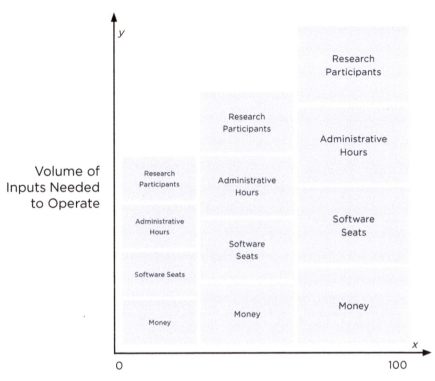

FIGURE 1.1

Proportions won't necessarily grow in nice tidy ratios, particularly if you find efficiencies as you scale, which you should. Even so, it's essential to work out the proportional volume in inputs that you'll need to sustain operations as demand grows.

Efficiency Does Not Guarantee Scalability

If you're a naturalist, you'll likely know that all toads are frogs, but not all frogs are toads. Similarly, while all scalable systems are efficient, not all efficient systems are scalable (see Figure 1.2). Say you deliver a service that can efficiently provide 50 researchers with access to research participants, but, as demand for the service grows, it slows down and becomes progressively less efficient. Although the service is (or was) efficient at a certain level, its efficiency doesn't scale, so it is not scalable.

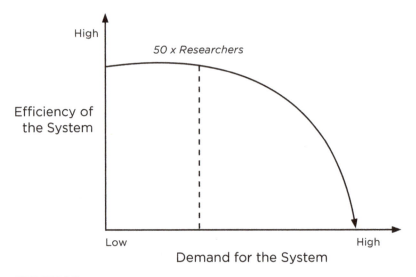

FIGURE 1.2
If the demand placed on an unscalable system exceeds an optimal point,
its efficiency (and effectiveness) will decline. Not every system needs to
be scalable, but you should always understand where the optimal point
is, which requires understanding the pressures, and proportions, of
supply and demand.

The key point to remember is this: a scalable system grows based
on predictable and proportional inputs and outputs, and it is *always*
efficient. Its predictability means that it retains the same levels of
efficiency and quality as demand ratchets up.

It Must Be Valuable

In a world obsessed with "more," it can seem that everything should
be built to scale. But it's not always necessary to make a system
scalable. It takes too much investment and smarts to make something
scalable, so why do it needlessly? Perhaps you deliver a scalable
system for storing research data, but if the quantity of data collected
is typically minimal or the number of people who access the system
can be counted on one hand, the scalable aspects of the system will
be pointless. So, before you decide to make a system scalable, consider whether the investment is necessary.

Finally, a system must be inherently valuable for it to be worth scaling. Perhaps you achieve the goal of giving 500 people access to research tools and training. Though this might be an operational triumph, if it doesn't ultimately improve the organization's understanding of its users or appreciation of good research, it may also be an inordinate operational misspend. "The definition of scalability depends on the *correctness* of a system, rather than the size or speed of it." Archis Gore, once again, hits the nail on the head.

Making decisions about what systems to scale to increase the value of research in the organization is a key part of defining a research and research operations strategy (see Chapter 2).

So, a scalable system is always:

- **Predictable and measured**—whether it's small, medium, or large does not matter.
- **Highly efficient**—efficiency alone doesn't guarantee that a system is scalable, but a scalable system is always highly efficient.
- **Necessary**—if the system handles minimal or unimportant demands, consider whether it's worth making it scalable at all.
- **Valuable**—unsatisfying experiences rarely stick around long enough to scale, so a system must deliver tangible value.

If you scale research with these truths in mind, you'll certainly be more successful than you might have been without them. But words like *predictable, proportional,* and *measured* hint at a tidy linearity that doesn't necessarily exist when it comes to operationalizing a knowledge trade like research. And though the word *scale* inspires comparisons with mass manufacturing, research doesn't evolve in straight lines or in nicely proportioned units, as easy as that would be, and original knowledge isn't (yet) crafted by machines.[3] Instead, scaling research needs an approach that is squiggly rather than boxy—communal rather than mechanical.

---

3 Generative AI tools like ChatGPT produce original content based on existing knowledge, but they aren't yet capable of producing new insights. When asked, ChatGPT replied that it is "designed to provide information and answer questions based on the data it has been trained on, up to its knowledge cutoff date. While it can generate responses and provide explanations based on that existing knowledge, it cannot produce new, original knowledge beyond what it has been trained on."

Research Is Not Mass Manufactured

When I was six years old, I went on a school trip to a bread-making factory where thousands of loaves of bread were produced in just one hour and conveyor belts and automated machines handled just about every part of the process. I'll never forget the enormity of the place, both in terms of its size and speed. At the end of the tour, we were all given a piping hot loaf of bread, and I have rarely been prouder of something so simple and so nonunique—every loaf a perfect clone.[4] This first memory of scaled-up production has become a personal meme for the mechanical repetitiveness that underpins scalability in all kinds of industries. Scalability relies on predictability and efficiency, which relies on big and valuable, words like *systematization, mechanization, standardization, repetition*, and *automation* (see Chapter 3, "From Strategy to Operational"). While a mechanical or linear approach to scaling is all well and good (if you're producing perfect clones), it doesn't work as well when your trade is knowledge. Knowledge isn't a static commodity: research outputs aren't factory-floor repeats, and research practices rarely grow linearly. Instead, they tend to grow not just in size but also in scope and stakeholder reach, which impacts scalability (see Figure 1.3).

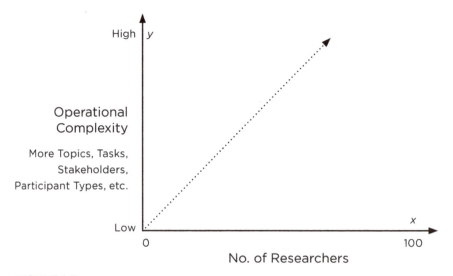

FIGURE 1.3

Typically, as the number of people doing research increases so does operational complexity. You must consider both the *x* and *y* axis when building and maintaining operations.

---

4 Unfortunately, the dog got hold of that loaf of bread, so I didn't enjoy even a bite!

Pretend for a moment that you're delivering operations for a team of ten researchers who research similar topics with similar participants using just two methodologies, perhaps remote moderated user interviews and usability testing. To set out operations, you might provide access to a standardized cohort of research participants, say people who regularly order takeout, and you provide access to the systems, protocols, training, and support that researchers need to do remote-moderated user interviews and usability testing.

The researchers are skilled, everything runs like clockwork (thanks in part to your operations), and the practice grows quickly. In fact, it triples in size. As a result, the research team now works on bigger remits across a growing product suite. They interrogate more diverse and complex research questions, and, to handle it all, they now employ a multitude of methodologies like ethnographic field trips, quantitative research, and brand and market research.

But every new area of focus, say a product or company priority, requires access to a unique set of participant cohorts, each of which requires different handling, and every research methodology requires unique operations, or at least a variation of them. Also, as the research team grows, they collect more and more data, and they create more and more content, so data is spread everywhere, and insights are increasingly harder to find. If you've worked in fast-growing research space, this will sound all too familiar.

While the original operations are scalable—they could scale endlessly—they aren't built to meet the increased mass and most important of all, the *variety* in researchers' logistical needs. The research team has scaled both up (in size) and out (in task variety), but its operations have only scaled up (see Figure 1.4). They haven't kept pace.

Countless research (and ResearchOps) teams bump into this trip hazard without even knowing it. At least, until it's too late. The solution lies less in a checklist of to-dos and more in shifting how you approach the work. These are the key points that you must keep in mind to avoid the same fate:

- **Things don't just scale, they scale multidimensionally.**
  Whether a practice is scaling up (more people are doing the same thing) or scaling out (the same number of people are doing a wider variety of things), or a mix of both (more people are doing more various things), it will significantly impact how you design

and deliver operations. Of course, practices can scale down, too, per the 2022/23 layoffs. (Hint: When it comes to scaling research, it is usually that more people are doing more various things.)

- **To sustainably scale research, operations must be scaled in tandem.** This doesn't mean that every time a researcher is hired, an operations headcount should be filled, too. It does mean that you must understand the proportions or ratios of investment that are required to sustain operations as both the scope and scale of your research practice grows. If you get greedy with resources and fail to simultaneously nurture operations, your research empire will fall. A lot of this understanding will come from tracking money and operating metrics (see Chapter 10, "Money and Metrics").

- **Finally, know that successful operations are *never* "set and forget."** Even the best-designed operations need constant investment, adjusting, and maturing to continue to meet demands. You'll learn more about this in Chapter 4, "Planning Realistic ResearchOps."

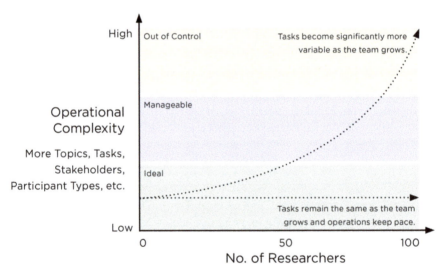

FIGURE 1.4
When an operating system isn't designed or adjusted to deal with increasing variability, it will tend to go into decline. This is especially true of highly variable tasks like participant recruitment.

While it's easy to jump to examples of mass manufacturing when you hear the word *scale*, the comparison isn't a perfect fit in the context of research. Though the basic tenets of standardization, automation, and mechanization hold true (see Chapter 3), there is a lot less to learn from mass production than might originally be assumed. But that doesn't mean that the mental models that are needed to scale research must be made up from scratch. There is a great deal to learn from a surprisingly tangential profession that works at large, and even super-large, scales. But instead of mechanized production lines, it involves communities of people, infrastructure, architecture, and culture. Sound familiar? It should because it is *just* like research.

Build a Research Metropolis

At a seminar in New Delhi in 1959, architect and author Laxman Mahadeo Chitale said in his speech that civic design is "The art that governs the siting and appearance of every material object from the lamp standard and the sidewalk to the community center and town hall, the parks and playgrounds necessary for the efficient functioning and progress of the modern urban community."[5] Although he said this 60 odd years ago, not a lot has changed. While there are countless analogies to explore between research operations and civic design—transit systems will be explored in a moment—it is the notion of a city, at least its aerial view, that provides a tangible map of what a mature research metropolis, or a *research operating system*, should include.

> **NOTE** A RESEARCH OPERATING SYSTEM
>
> In her book, *Scaling People: Tactics for Management and Company Building*, Claire Hughes Johnson wrote, "An operating system is a set of norms and actions that are shared with everyone in the company. These shared systems and parameters are essential to growth and success." Likewise, a research operating system is a consistent and shared set of infrastructure, norms, and actions that help people to consistently deliver, take part in, and gain value from research.

---

5   L. M. Chltale, "Civic Design," in *Seminar on Architecture*, edited by Achyut Kanvinde, 182–192. New Delhi: Lalit Kala Akademi, 1959, https://architexturez. net/doc/az-cf-168640

A mature research operating system is eerily similar to a well-planned city (see Figure 1.5), even if it's typically built in binary code as opposed to bricks and mortar—unless you're building something physical like a research lab, of course. A city needs spaces for people to connect, a library to access knowledge, services for accessing support, and signage to know where to go. It also needs ways to "keep the lights on" with civic services like sanitization (or data hygiene), telecommunications, and more. It needs a judicial system to ensure respect and privacy, a financial system to keep money moving, and spaces and etiquette for working well together. It needs models for measuring traffic and population growth to support planning, and it must engage with organizations both nationally and internationally to both operate and grow. Finally, it will need to provide ways for people to learn new skills and be inspired to do more. These are the fundamental requirements of all progressive metropolises, and they are the basic requirements of progressive research operating systems, too. These basic requirements are defined in this book as the *eight elements of ResearchOps*, which will be explored in Chapter 4.

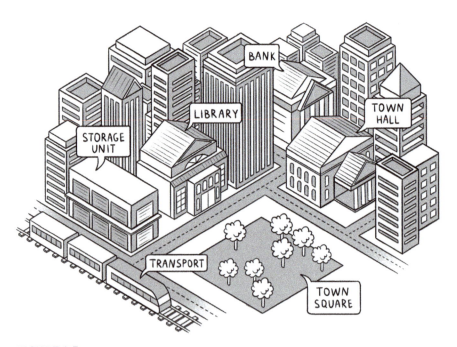

FIGURE 1.5
The "skyline" of a mature research metropolis is remarkably like a well-developed city.

This zoomed-out view of the major parts of your operating system is essential for designing a cohesive and complete "research metropolis," but treating common ResearchOps features like a library, panel, or field-research kit—or even worse, aspects or elements of ResearchOps like participant recruitment, tooling, or knowledge management—as if they were a checklist of siloed things will *not* scale research. A city only thrives because people can get to where they need to go, and they succeed when they get there. Even the most impressive cities will drive people out if they make life too hard, or hamper peoples' movement, safety, and success. If you've ever lived in San Francisco (or my hometown, Johannesburg), you will know what I mean: easy transit is everything.

The Beauty of Transit Systems

One of the world's most iconic transit systems is the London Underground, better known as the *Tube*. The Tube is iconic not just because it is in London, but because, considering its complexity and size, it is incredibly efficient and easy to navigate. The Tube shuffles up to three million passengers around 272 stations across the capital each day,[6] boggling numbers by anyone's standards, and more than 543 trains zip around its network at peak times. *That's* scale!

The kind of system that drives the London Underground is far more analogous to building scalable research operating systems than the linear schemata of mass production (see Figure 1.6). It still uses standardization, automation, and mechanization in the form of tracks, trains, signaling, signage, timetables, and more, but rather than restricting people to a limited set of A-to-B journeys, it gives people everything they need to independently get to where they need to be. And though it is highly systematic, it empowers millions of people to independently travel countless unique journeys. *This* is the key to scaling research.

---

6   Transport for London, *What We Do*, https://tfl.gov.uk/corporate/about-tfl/what-we-do

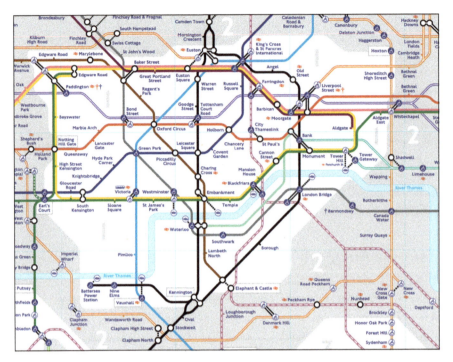

FIGURE 1.6

The iconic London Underground map helps millions of people navigate a complex system to get where they need to be fairly easily—if not always speedily.

If you look at a map of the Tube, or the map of any transit system, you will notice that it is really a kind of decision tree (see Figure 1.6). To stick with London, if your starting point is Waterloo station and you want to get to Hyde Park Corner, you could get there in multiple ways depending on your needs and priorities. Using the Transport for London app, it takes moments to discover that the quickest route is a 26-minute trip involving two trains, one change, and a six-minute walk (see Figure 1.7). And if you start the journey, then change your plans halfway, you can hop on and off, and alter your route without too much hassle. In essence, the Tube makes moving between the capital's hotspots incredibly easy; your research operating system should do the same.

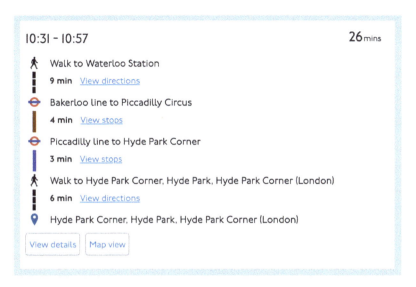

FIGURE 1.7
Transit and other civic service websites offer great inspiration for how you might support users to navigate a complex research operating system; even the best-designed systems can be complex.

Research Logistics Are Decision Trees

While it's common to present arranging research as a series of repetitive steps—plan, recruit, schedule, consent, do, analyze, document, present—just as research teams don't scale linearly, the logistics involved in running a research study typically aren't linear either. While some researchers might have some degree of logistical continuity, they frequently need to meet different needs as they move from study to study, and often within a study.

In one study, a researcher might need to hone the research question with stakeholders, so they do a literature review of existing research. Then they need to recruit two participant cohorts, say hard-to-reach individuals to take part in user interviews, and a subset of customers to take part in a diary study. They might work with a recruitment agency to access the hard-to-reach people. To access the subset of customers, they might give the in-house customer panel or product intercepts a go; the latter tends to require engaging with product owners. If recruitment proves tricky, they may need to work with an analyst or sales or support partner to access a list of customers, which isn't always easy.

To arrange the study, the researcher will have to traverse a multitude of partnerships, platforms, and tools, and they may need to hop on and off at certain points or change tack entirely. They'll also need to make sure that their communication styles, consent agreements, thank-you gifts, and more, are audience appropriate (somewhat like knowing the culture and etiquette of the area you are visiting).

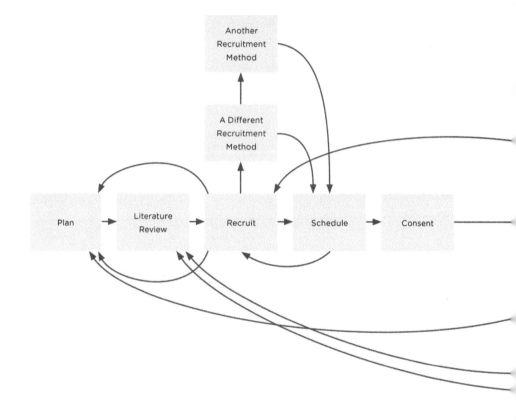

If you're a researcher, I imagine that you're feeling heard and seen right now. It's a lot to take on! And it is certainly not linear (see Figure 1.8). Also, other people like research stakeholders, consumers, vendors, and operations will travel the same journeys, and intersect at points with other people, so the system must cater equally to their needs and experiences.

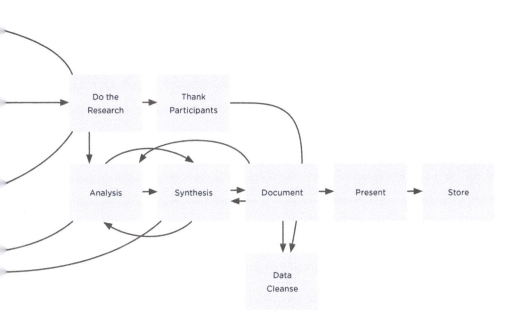

FIGURE 1.8
While simplistic diagrams of the research process are convenient, they don't represent the reality. Unless everything runs perfectly according to plan, or momentum and curiosity are low, researchers will often use multiple avenues to recruit the right participants, or they may adapt in response to new knowledge or do a literature review to interrogate insights halfway through.

The transit system analogy will break if you push it too far, but there are several important concepts that ResearchOps can practically apply. To deliver a scalable research operating system you must do the following:

- **Consolidate protocols and tools under "stations."** A *station* should be context-orientated rather than focused on a task, like recruitment, or a vendor brand name, like UserTesting.[7] If the destination is "accessibility research," you might provide access to agencies that specialize in recruiting people who are living with disabilities, access to sign-language interpreters, specialist training, and advice on appropriate tools to use. Remember that delivering a list of resources is not a system. A system must be complete, interconnected, and navigable—just like the Tube.

- **Surface as much of the system as needed.** Share just enough about how the system works to help people understand and navigate it. You can do this by providing a literal map of your services, like a decision tree. You might also provide estimates on how long it might take to get something done, what to do if something doesn't work, or a dashboard that gives an overview of wait times and system outages.

- **Be transparent about your system's limits.** Let people know what they should do if they run out of "track." If there are no standardized resources and protocols to support their particular journey or need, make it clear as to what they should—and should *not*—do to complete their journey successfully.

- **Provide a full-service option to handle sensitive needs.** Somewhat like a taxi or chauffeur service, it's important that your system includes high-support services to handle logistics when a study is time-sensitive, ethically sensitive, or if it involves VIPs. Though public transport systems are highly scalable, a packed train isn't always an appropriate experience. Standard operating models will be covered in Chapter 3.

- **Be aware of how your operating system interacts with other systems.** The Tube is part of a wider transit network that is run by Transport for London, which connects with a variety of other systems like a bus and national rail networks. Research operating systems are no different: they must comply with external regulations, particularly to do with privacy, and dovetail with the wider organization's operations.

---

7 UserTesting, www.usertesting.com

- **Deliver operations piece by piece, but in line with a master plan.** The Tube opened on January 10, 1863, with just one line and six stations, now there are 272 stations to help people navigate a fast-paced 21st century city above and below ground. Delivering operations piece by piece, but with a larger system firmly in mind is the reality of scaling complex research systems, and it is key to strategy.

Everything from transit systems to stadiums and airports, and entire cities, contain near infinite inspiration for nurturing the mindset that's needed to design scalable ResearchOps systems. But this isn't just because research operating systems are large infrastructural systems that are built for people; they also have the power to change how people think, feel, and behave. This is a crucial point. The success of research rests on knowledge exchange, which rests on *trust*, which rests on culture. So, if the culture—the biases, values, and attitudes—that surround research isn't ideal, you will find it hard, even impossible, to scale research however well set up your research operations.

Scaling Is *All* About Culture

Years ago, I watched a documentary about ancient Rome, or perhaps it was about capitalism, I can no longer recall, and it opened with a concept that has stuck with me: Rome was successful in its conquests not only because it had a strong army, but because it brought to its conquered nations a culture that was revered and even desired. It's a tricky business referencing war, and I am no historian, but disclaimers aside, the story illustrates an important point: to scale research, you will need to do more than assemble a legion of researchers with best-in-class infrastructure. In addition to this, you must consider how your research operating system, and each part that comprises it, will impact (and even mold) the culture that surrounds research— how it will make people *feel*.

Culture is another big and bandied-about word in the business world. Culture is defined by what a group of people, like a team or an entire company, value—and do not value—along with their beliefs, biases, and attitudes. I use the term *research culture* to mean the values, beliefs, biases, and attitudes that people have about what good research looks like, how it should be executed, and the value that it should provide to them. Culture is crucial to the success of a research practice because trust and receptivity are intrinsic to people accepting new knowledge, and successful knowledge transfer, or learning, defines the value of research.

Pulling Cultural Levers

Several years ago, while working at a large company, I was asked to fix participant recruitment, a common request of ResearchOps. In response, I hired an experienced recruiter, built a service desk, then offered full-service participant recruitment to everyone in the company—for free. Whoever you were, you could submit a ticket to the service desk and within two weeks and almost no effort, you would be meeting with an ideal set of customers. The dream! (And naiveté, as it turned out.)

Even though it seemed like a good idea at the time, near-painless access to participants made connecting with customers feel cheap, and it led to research that was hastily planned and executed. Instead of bolstering customer understanding, the effort downgraded the quality of research being done by nonresearchers, and it negatively impacted the appreciation of research as a craft. It didn't take long before we shut the service down; not to give up but to create operations that provoked a better research culture. A year later and armed with these learnings, we once again gave product teams access to participants. But this time via a self-service recruitment platform, and, instead of unlimited access, each team had a monthly participant budget, and we provided training on how to recruit well. This trifecta—technology, constraint, and skills—encouraged product teams to carefully work out their research priorities, put effort into conducting the research, and consolidate research efforts across teams, which encouraged further collaborations. These were all great knock-on effects of the research operations.

When you design your operating systems, make conscientious decisions not just about *what* systems you'll deliver, but *how* you'll deliver them so that they proactively shape the values, biases, and attitudes around research for the better. Your operations can either slow people down or speed them up, or make research feel accessible or out-of-reach, expensive or cheap. It's not that one is always better than the other, sometimes there's a benefit in making things hard to access, say in the case of customer data, or slowing people down, say to properly analyze data or to be present when observing research. The experience that your operating system delivers (the entire system and each of its parts) will shape people's values, biases, attitudes, and expectations around research. The shared culture that this creates will impact, for better or worse, the quality of the research that's generated and how it's perceived, which will ultimately impact scalability.

In a Nutshell

To scale research, and to deliver research operations, you must deliver more than a disconnected or randomly appended collection of research tools, training, and assets. Instead, you will need to create a research operating system: an interconnected and standardized set of assets and actions that are shared (and trusted) by everyone who is involved in research. To succeed, your operating system must be the following:

- **Complete:** It must include all the necessary systems that are needed to operate research on a day-to-day basis (see Chapter 4).

- **Commutable:** Arranging the logistics of research is rarely a straightforward journey from point A to B, so people should be able to easily navigate a million different research decision trees (see Chapter 7, "Seamless Support").

- **Cultural:** Your operating system as a whole, and every system within it, will create a vibe or a spirit. Rather than simply decide *what* you will deliver, you must decide *how* you will deliver operations to shape your stakeholders' values, biases, and expectations about research. The research culture is the biggest indicator of how successful your research practice will be.

Opportunities to scale the value of research are literally endless, and there are seldom enough resources available to do it all. But the good news is that you don't have to do it all—in fact, doing it all is impossible. Instead, you will need to define a discrete set of high-priority goals that have clarity and vision, and realistic plans around how you will operate to achieve them. A research strategy should outline *what* should be achieved; a research operations strategy should outline *how* it will be achieved, and the two should be in perfect synchrony.

CHAPTER 2

Lost and Won on Strategy

If you've ever watched Formula 1 motor racing, you'll know that it's a highly strategic sport. Even if a team has the best cars and two talented drivers, if they don't set their cars up to suit the track and the weather on the day, and if they don't get the timing and management of everything just right, from pit stops to tires and batteries to race pace, they're unlikely to end the day with a trophy. But to win races, teams must plan well beyond each race. They must have detailed strategies for developing drivers and cars over multiple years, and they must know how they'll approach each racing season, each of its 24[1]-race weekends, and each part of a complex system for making cars go incredibly fast. And when the rubber hits the tarmac, they must adapt well-made plans, sometimes within a split second, to make the most of what is happening right there and then: if a safety car is deployed in response to a crash, a well-chosen pit stop can make or break a driver's chance of being on the podium.

The notion that strategy is essential to running a system is not new, whether the system is an F1 team, a driving machine, a business, or a research practice and its operating system. But that's not to say that strategy is typically well understood. In fact, there are few words that are as widely used yet cause so much angst and confusion. Because *good* strategy is so crucial to delivering a scalable research operating system, it's vital that research (and ResearchOps) leaders embrace it. This chapter uncovers the status of strategy within the research profession, and why it's essential to shift it. It tackles the concepts of value and trust in research, and how a *research strategy*, and its perfect pair, a *ResearchOps strategy*, should dovetail to make sure that dollars spent deliver perceivable value to the business.

> **NOTE** RESEARCH STRATEGY VERSUS STRATEGIC RESEARCH
>
> In recent years, there's been a lot of conversation about *strategic research*, which is not to be confused with *research strategy*. Strategic research is research that informs strategic decisions about products or services. A research strategy defines what a research practice will do—and *not* do—to deliver value to the organization. It should guide everything from the structure of the research team to day-to-day decisions, and it should inform a ResearchOps strategy, too. It's highly inefficient (and plain ineffective) to set up operations if strategic goals are nonexistent or murky.

---

1   The 2024 Formula 1 calendar comprises a record-breaking 24 Grands Prix.

Strategy *Is* the Blind Spot

In early 2024, I ran a LinkedIn poll[2] to understand how many research teams were operating with or without a research strategy. In the poll, 166 researchers and people in related fields took the time to respond, and the results were sobering: 55% of the respondents' teams did not have a research strategy while 10% were unsure, which is as good as not having a strategy at all. This means that just 35% of respondents' teams reported having a research-specific strategy to guide their work. More importantly, 65% did not.

It was a simple poll, but it showed that there is a crucial gap in a profession that in recent times has seen significant cuts to its workforce. User Interviews' "The State of User Research 2023" report noted that "Half of researchers were directly or indirectly affected by layoffs in the last 12 months,"[3] and there were more layoffs to come. As I'm sure you'll agree, it's a statistic that relates to a highly specific point in time—a black swan event,[4] even—but it offers evergreen lessons for how to build a research practice that's resilient when times are tough and that thrives when times are good. Because, if one thing has been made clear, it's that research is not immune to the standard business math of return on investment or cost versus value. And why would it be? It is also clear that good business management, though it might seem less gutsy or visionary (or user-centered) at the time, will always outlast the "we-are-invincible-type" spending that tends to come with an economic boom—or a research boom.

The Big Research Boom

Around October 2022, the world's biggest tech companies reported earnings of billions of dollars that were wiped off their value suddenly. So ended a tech boom of 15 years that had set the scene for many disciplines, including user research and research operations, to thrive as never before. For example, fifteen years ago, in-house research teams of any size were a rarity, never mind teams that run into the dozens and even hundreds as is more common today

---

2   View the results of the LinkedIn poll: www.linkedin.com/posts/katetowsey_userresearch-researchops-activity-7153979808769609729-KvCq/

3   "The State of User Research 2023," User Interviews, www.userinterviews.com/state-of-user-research-2023-report

4   According to Britannica, a black swan event is "a high-impact event that is difficult to predict under normal circumstances but that in retrospect appears to have been inevitable," www.britannica.com/topic/black-swan-event

(and include a team dedicated to ops). More often than not, these teams had been formed under the general belief that understanding customers was good and that a band of researchers spread around the organization could deliver all the understanding required to create better customer experiences. And that ops could help them work more efficiently. But it's a distributed *operating model* that hasn't always worked. As with any other part of a business—marketing, finances, IT, and so on—research leaders need to do more than just hire more people to do more of the same work, i.e., more research, at an ever-increasing cost to the business. They must also have a practical and prioritized plan (a strategy) for how to use limited resources, and resources are *always* limited, to contribute value where it matters most, to whom it matters most, and, truth be told, sometimes where it will most be seen.

ONE THING GOES BUST, ANOTHER GOES BOOM

It's true that user research professionals of all types took a post-pandemic hit (ops, contractors, and agencies, should not be forgotten); however, the notion that user research is required to build successful products has *not* been disputed. Instead, the loci of investment has shifted from building specialist research teams to empowering (and often simply offloading) the job of research onto other disciplines, i.e., "democratizing" research. As a result, tools that promise "research in hours, not days" have survived relatively unscathed, and practices that provide nonresearchers with a framework for spending time with customers—cue Teresa Torres's *continuous discovery*[5]—have been widely adopted.

In the long term, this trend doesn't mean that bona fide researchers are out of a job. Instead, by empowering everyone to spend time with customers in ways that suit their skillset and context, researchers' specialist expertise can be utilized to answer more high-value or complex research questions. Also, research leaders can manage more laterally (see "When Everything Looks Like a Nail"), and ResearchOps can provide the systems that underpin all manner of user, product, and sector-oriented learning. As ever, there's opportunity in change (if you are willing to adapt).

---

5 In 2022–2023, the continuous discovery approach to understanding customers took the product world by storm. This was fuelled by Teresa Torres's bestselling book, *Continuous Discovery Habits: Discover Products That Create Customer Value and Business Value.*

Research strategist Chris Geison said in a blog post, "We talk a lot in this field about *how* to do research, but not enough about what research should be done." He added that teams that have a research strategy "are contributing meaningfully to the mission of their organization. They're also prioritizing effectively, allocating resources effectively, and they're not wasting money on projects they shouldn't do. Nor are they wasting researchers' time on research that won't get used."[6] This is despite the garrison of researchers who have been hired into senior management roles as part of the boom: Head of User Research, Director of Customer Research, Senior Research Manager, and VP of Research and Design are a small sampling of the top-tier roles that are more common today.

Whether explicit or not, these researchers have been asked to step up and become, not just research leaders, but *business* leaders: the job of running a research practice is as much about good business management as it is about research expertise. Research leaders must be well-informed and well-networked, strategic about how they model their teams and structure their work, i.e., how research operates, an arch people leader because an invigorated pool of talent is key, they must deliver measurable results that benefit organizational priorities and balance the insatiable need for insights at speed with scaled-up and reliable research, and, finally, they must understand how the business works, not to mention its politics. It's not an insignificant list, but, if you're leading a team of research professionals, which includes research operations, *that* is the brief. And none of this can be done without a research-specific strategy.

Before diving headlong into the ins-and-outs of research strategy, it's worth unpacking what makes a *good* strategy good, regardless of its departmental focus. Because strategy isn't all about "futurizing" and ambition.

---

6 Ben Wiedmaier, "Contribute Meaningfully: The Power of a Research Strategy," *dscout* (conversations), https://dscout.com/people-nerds/geison-research-strategy

Strategy Is Not Futurist Fantasy and Fluff

Strategy is often associated with big ideas, aha! moments, and late-night genius, but a good strategy is nothing more than a clear, logical, and prioritized plan in response to well-defined challenges or opportunities. It needn't be visionary, bold, or drawn-out work. It's this simple set of misassumptions that often paralyzes smart people confronted with defining a strategy, something I've experienced first-hand. I once procrastinated producing a ResearchOps strategy for over a year because I assumed that I needed more time, information, and creativity to do the job when I simply needed to face up to the challenges at hand. Then it's not uncommon to come across confident strategists who miss the salient points of what makes a good strategy good in lieu of mindboggling concepts that are hard to understand and even harder to execute, which isn't the point at all.

In his highly recommended book, *Good Strategy Bad Strategy*, Richard Rumelt wrote: "A good strategy does more than urge us forward toward a goal or vision. A good strategy honestly acknowledges the challenges being faced and provides an approach to overcoming them."[7] A good strategy should be:

- **Prioritized:** Deciding what to do—and what not to do—is the heart of good strategy. A stack-ranked list of three to four priorities is ideal. Even huge companies like Apple and Disney focus on a list of corporate strategic priorities shorter than the number of fingers on one hand.

- **Actionable:** A strategy is a practical plan of action. It should do more than espouse lofty values, visions, and missions, as heartwarming as those are. You might consider using SMART goals—specific (see below), measurable, achievable, relevant, and time-bound[8]—to define strategic priorities so that they're actionable.

- **Specific:** Priorities should address specific challenges, opportunities, stakeholders, goals, and a specific time period, etc. Is the strategy relevant to two weeks, one month, or several years?

---

7   Richard Rumelt, *Good Strategy Bad Strategy: The Difference and Why It Matters*, (Great Britain: Profile Books Ltd., 2017), 4.

8   Kat Boogaard, "How to Write Smart Goals," *Atlassian* (blog), December 26, 2023, www.atlassian.com/blog/productivity/how-to-write-smart-goals

Is it relevant to these stakeholders, but not those? Or only these types of research methodologies?

- **Informing and responding:** A good strategy never exists in a silo. It should respond to the wider environment, and it should inform and respond to the strategies that are superior and subordinate to it.

- **Committed:** It takes boldness to turn theory into reality because it usually means change. So, commit to seeing a strategy through and keep in mind that results often aren't instant.

- **Evolving:** Commitment shouldn't be confused with statis. A good strategy should shift in response to change; it should be directive (guided from the top) and emergent (informed bottom-up by new learnings and evolving realities).

- **Simple:** A strategy should be to-the-point and easy-to-understand. The best strategies are often just a few concise paragraphs (one for each priority) and a table defining how, who, what, where, and when. Save poignant writing skills for philosophy and poetry!

- **Shared:** It should get everyone and everything working in synchrony to achieve the same goals. As the CEO of Disney Bob Iger said, "Your strategy is only as good as your ability to articulate it."

The word *strategy* has been a part of common language since the early 19th century, and it has practical military roots. So, if you find yourself held back by the modern notion that a strategy is a fancy document bursting with innovative ideas, remember its wartime roots. As Rumelt says, "The successful strategists ask, 'What's the crux of these problems? Can I get through them? And if so, which of these problems is worth putting our resources toward?'"[9]

FOOD FOR THOUGHT

STRATEGY IS A NOUN AND A VERB

This concept is mentioned a few times in this chapter, but it's so important that it deserves repeating. A strategy should be codified in a document so that the plan can be referred back to, progress can be tracked, and to get everyone on the same page, it should be a noun. But strategy is more about *doing* than documenting. Once you know what must be achieved and how

---

9   Yuval Atsmon, "Why Bad Strategy Is a 'Social Contagion,'" *McKinsey & Company* (podcast transcript), November 2, 2022, www.mckinsey.com/capabilities/strategy-and-corporate-finance/our-insights/why-bad-strategy-is-a-social-contagion

to get there, you'll need to make constant smart and logical decisions, say about operating systems and subsystems and people required to achieve strategic goals.

Why Research Needs Its Own Strategy

In a conversation about strategy, a senior research leader said: "But we follow the product team's strategy. Why would we need a research strategy?" It's often assumed that the company's strategy or the strategies of key stakeholders are enough to guide a research practice, but these strategies don't offer a consolidated point of view about where and how the research practice should focus *its* efforts and unique capabilities to deliver value—you could think of this as the unique selling point (USP) or competitive advantage of the research team, which is food for thought if research is being democratized around you.

In an article published by *Harvard Business Review*, differentiation strategist Andrea Belk Olson wrote: "Many leaders would argue that there should be only one strategy for a company, and that strategy provides all the guidance and direction necessary for departments to create implementation plans. However, most high-level strategies have hundreds of ways they can be translated into action."[10] Research leaders aren't alone in thinking that adopting another department's strategy, or the entire organization's strategy, is good enough. Olson says, "A corporate strategy is intended to focus an organization on what should and should not be done to succeed—a department supporting strategy should do the same."

So, a research strategy should outline how the research practice will support the organization to achieve its goals. Say you work for a music streaming business and the CEO has recently announced a strategy comprised of four priorities. One of the priorities is to expand into five new markets: India, Japan, Brazil, Mexico, and Sweden. Research could do a lot to support this goal—the economics and cultures of these markets are quite different—so you give it top spot in your research strategy and come up with a viable plan for how to operate to achieve the goal (see Chapter 3, "From Strategy to Operational").

---

10 Andrea Belk Olson, "3 Reasons Why Every Department Needs Its Own Strategy," *Harvard Business Review*, January 26, 2024, https://hbr.org/2024/01/3-reasons-why-every-department-needs-its-own-strategy

By definition, a corporate strategy outlines what executive management values the most and where they'll invest money, attention, and time, so it makes sense to align. But that's not to say that the corporate strategy should be the sole informant of the research strategy—reality is rarely that simple or ideal.

Everything You Could, Should, and *Must* Do

Just as the executive team will have analyzed multiple factors—market forces, competitor moves, big ticket customers' needs, board requests, company culture, and new technologies—and placed bets on a limited list of initiatives primed to support sales or savings, a research leader must do the same. When devising a research strategy, and in addition to the corporate strategy, you must consider a kaleidoscope of factors like:

- Stakeholder priorities that didn't make it into the corporate strategy, but which are still important, viable, and align with the unique capabilities of research.

- Practical efforts for maturing how research is perceived and valued within the organization.

- Opportunities for operational improvement to widen your reach, add value, or make savings.

- How research might leverage new technologies. The timeliest example is artificial intelligence (AI), but new capabilities pop up all the time.

- The innate strengths of the team and how you might use them to your advantage (or train or hire to fill gaps).

- Challenges that stand between you and delivering value. These may even be things that irk or frustrate! Bah, humbug to the idea that business should be unemotional; just make sure to ground responses with pragmatism and logic.

- Crucial factors that the executive team are missing or need to better understand. If you choose to take this influential route, you'll need to secure buy-in first. (See "Building a Foundation of Trust.")

- Events happening in the wider world that aren't explicit in the corporate strategy, like a company merger, a shortage of researchers for hire, or a global economic shake down (or uptick).

If you have some sort of notetaking gadgetry nearby, take a moment to list the things that come to mind when reading the above list.

Without too much effort, you'll likely be able to list at least 10 points; there's rarely a shortage of things to do! But a strategy is defined as much by what you do as what you do *not* do, which requires ruthless prioritization and the commitment to see priorities through.

> **FOOD FOR THOUGHT**
> **YOU CAN'T HAVE YOUR CAKE AND EAT IT, TOO**
>
> It's not unusual to come across teams that have a strategy, but they're expected to achieve the goals in addition to work that's already going on. In other words, the strategy is less of a plan for how to operate to achieve high-value goals, and more of a laundry list of side projects. The definition of a priority is that it carries more weight (because it will deliver more value) than *all* other work. This means that your team must be empowered with the structure, time, skills, tools, and network they need to focus on this work above all else, which often means committing—and committing to change.

Defining a Research Strategy

Sometime in the mid noughties, I played double bass in an orchestra. A full-scale symphony orchestra has around 100 musicians playing at least 15 different types of instruments, from violins to trumpets and kettledrums to double bass. While it's the job of the conductor to lead the orchestra, each section of instruments—I was one of three bassists that made up the bass section—must follow section-specific sheet music to play the right sounds at the right time. "Sounds" is the right word to use because, particularly as a bassist, it's rare to play a melody. Instead, you might add a long "bwaaaaaaah" here and a short set of "bah, bah, baaaahs" there. Solo, the bass part sounds disconnected and unimpressive, but in coordination with all the other instruments, the result is *music*.

Strategy is teamwork and coordination at its height, but it takes knowing what to do—and what *not* do—with your particular instrument to add value that's in tune with the rest of the organization and the audience.

At its simplest, producing a strategy (as a noun: a thing) has three primary steps: you must choose a limited list of priorities, devise strategies for how to achieve each priority, and then write it all down to create a document of some sort. But a strategy is only as good as its execution, so, as mentioned earlier, you should also treat it as a verb (a constant doing), which requires additional steps. The full set of steps are as follows:

1. **List challenges and opportunities.** The first step to prioritizing is knowing everything that you could choose to do so that you can make smart choices about what to do, and what not to do. As explained under "Everything You Could, Should, and *Must* Do," you should take everything (impactful) into account that is going on within and around the research practice.

2. **Choose up to four priorities.** Priorities should deliver tangible value to the business, either immediately or later on down the line, and they should be practical considering resources. Read Harry Max's book *Managing Priorities: How to Create Better Plans and Make Smarter Decisions* (Two Waves Books, 2024); prioritization is a skill unto itself.

3. **Define a strategy for achieving each priority.** A strategy is more than just a list of priorities, you must outline how you'll achieve your goals. Make sure that your strategies aren't knee-jerk reactions or operational tropes, like building a research library or a recruitment panel without a well-defined goal.

4. **Get everybody onboard.** Wherever you sit in the organization, you'll need to collaborate with people above, to the side, and below you, which might require selling them the idea. I once delivered a strategy called "go/customer-centric" and presented it to 25 design teams. It meant doing the same presentation 25 times in just a few weeks—I was talked out! But it paid off.

5. **Make priorities operational.** This is the most important step, and yet it's also the most commonly forgotten. A strategy will only succeed if it becomes part and parcel of how the organization operates. This crucial topic is the heart and soul of Chapter 3.

6. **Measure the results and communicate.** Measuring the impact of research is tricky, but if you acknowledge that research does not scale, systems do (systems are easy to measure), and you have a clear view of how research is valued in the business context, it is less tricky than you might think. (More under "Forget About the Bottom Line.")

7. **Iterate.** You'll learn as you go and things will inevitably change, so make sure to adjust strategies or tactics in response. Just avoid the temptation to chop and change because unsubstantiated doubts have crept in, or you're being lured away by something seemingly more exciting. Anxiety and boredom are poor reasons for upending a strategic direction.

You Need a ResearchOps Strategy, Too

The steps required to define a research strategy aren't unique from those needed to define any other kind of business strategy, and the same rules apply with a ResearchOps strategy. A ResearchOps strategy should outline how ops will support the research practice to achieve its goals and what it must do from an operations point of view to deliver value and save costs. The research strategy and the ops strategy should ideally be in perfect sync, like Ginger Rogers and Fred Astaire. And if the research practice doesn't have a research strategy, devise a ResearchOps strategy anyway. Operations take too much investment and time not to know where you're going, when, and why. Besides operations is systems and all good systems are specific, which requires the specificity of strategic priorities.

> **FOOD FOR THOUGHT**
> **MAKING THE REALITY YOU WANT**
>
> If researchers are working within product teams across the organization or responding piecemeal to research requests (a distributed operating model), and they're committed to continuing that work, you might rightly argue that there is no centralized team to shift in response to corporate strategies. There are countless challenges that might stand in the way of doing what is ideal: a bad management structure, poor strategic skills at the executive level, too little buy-in, stakeholders who don't understand research, and more. But it is precisely these sorts of challenges that a research strategy should address. A strategy is a logical and prioritized plan, and its raw inspiration is the annoyances and frustrations that hinder the delivery of business value.

The content shared in this chapter thus far has been a level-set; it's nothing extraordinary to research. Research strategy doesn't require new tools or innovative ideas, and it doesn't require that you have an MBA. In fact, the advice shared in countless books and articles about business strategies apply to research strategies, too. I've already mentioned *Good Strategy Bad Strategy* by Richard Rumelt, but you should also read his more recent book, *The Crux: How Leaders Become Strategists*. And *HBR's 10 Must Reads on Strategy* by Harvard Business Review is a great addition to your library.

Although a research strategy isn't unique, there are strategic approaches that are specific to scaling research that you should keep front of mind as you devise a research strategy, and then get on with it. First, to deliver perceivable value to the business, you must understand how the value of research is perceived by the business; it's not tied to the bottom line. Second, whatever your scale, scope, and modus operandi, you'll need a plan for how to rise above the problem of value distributed piecemeal across stakeholders—a good problem to have—to delivering value that's unmistakable at the level of the executive board and corporate general ledger. Finally, no matter how reliable your research outputs or emboldened your plans, you'll fail to deliver value if research consumers aren't primed to listen to research, which relies on influence and trust.

Forget About the Bottom Line

In 1945, the management thought-leader Peter Drucker coined the term profit center. He used it to describe the divisions of an organization—say sales, product, or marketing—that are expected to directly add to the company's bottom line. The opposite of a profit center is a cost center. Cost centers are expected to provide the essential services needed to run a business, but they are *not* expected to be able to tie their value to the company's income: IT, human resources, and research are examples of cost centers. It's said that Drucker regretted this theory and later recanted it when he wrote, "Inside an organization, there are only cost centers. The only profit center is a customer whose check has not bounced."[11] Even so, the notion of profit versus cost centers has stuck, and it has formed the basis for more complex frameworks for measuring the value that a department is expected to provide. Included in this is the notion that cost centers might transform themselves through strategy and operational efficiency into a value center: a nonrevenue-generating function that delivers measurable value to the organization, which is a useful notion to lean into when it comes to managing a research practice.

It's not uncommon to read a post on LinkedIn or *Medium* about measuring the impact that research has had on the business's bottom line, the goal being to quantify (and sometimes qualify) the value of research. But it is rarely possible to make a direct, or accurate, or believable link between a research effort and a pivotal business metric like customer growth or churn, and there are several reasons

---

11 Peter F. Drucker, "The Information Executives Truly Need," *Harvard Business Review*, January–February 1995, https://hbr.org/1995/01/the-information-executives-truly-need

for this. For one, these kinds of business metrics are often the result of countless teams' efforts, so it's difficult, if not impossible, to neatly parse out the contribution of research whether as an ongoing engagement or an individual report. While a sales team can closely track the income generated by individual salespeople—their salaries are even based on their results—research is not as straightforward. Though someone might be influenced by reading a single report, their decisions are more likely to be influenced by a combination of exchanges like several research reports, a presentation, a chat with a researcher, observing a session or two, and then, mixed in with their existing biases, attitudes, and experiences, they have an aha! moment or a slowly matured understanding. Then one day, a proverbial lightbulb goes on. Even if the realization isn't associated with the work of the research team.

> **NOTE** **THE CALLS, CLICKS, AND CHURN OF RESEARCH**
>
> While you can't easily measure the point or gravity of research impact, you can measure engagement with your operating systems: the calls, clicks, and churn of research. For example, the number of people who attend research presentations, who take the time to observe research sessions, recruit participants, or the level of engagement with types of research content. (More on this in Chapter 10, "Money and Metrics.")

Outside of the difficulty of accurately measuring the impact of research, there's an even more important point: It's widely accepted in business management that research is a cost center. While a research department is expected to operate within a prenegotiated budget, which must be used to deliver tangible value, it is not expected to be able to tie its activities to the bottom line. In other words, efforts to do this are neither accurate nor required, so don't waste your time.

In his working paper "The Demise of Cost and Profit Centers,"[12] Harvard Business School emeritus professor Robert Kaplan

---

12 Robert Kaplan, "The Demise of Cost and Profit Centers" (working paper, Harvard Business School, 2006), www.hbs.edu/ris/Publication%20Files/07-030.pdf

explained that financial metrics are an incomplete measure for all types of business units whatever their label. He added that, whatever the context, measuring value should always include:

> ... a variety of nonfinancial metrics designed to capture how well the unit's intangible assets built and expanded relationships with targeted customers; improved the quality and responsiveness of operating processes; created and introduced new products and services; enhanced the motivation and capabilities of employees; leveraged investments in data bases and information technology; and aligned the culture and climate linked to the unit's vision, mission, and strategy.

It is a fantastic list. *This* is how research leaders, including ops, should be seeking to add, measure, and prove the value of research, not by chasing a fiscal line. Also, Kaplan's quote inspires a vastly more varied approach to delivering value than hiring more researchers to do ever greater amounts of research, which can feel like growth—but is it?

When Everything Looks Like a Nail

The American philosopher, Abraham Maslow, wrote in 1966, "If the only tool you have is a hammer, it is tempting to treat everything as if it were a nail."[13] Likewise, if your job is to do research, you're likely to solve problems by doing more research. If you're a research manager, you may default to hiring more researchers. If you work in operations, you may want to turn everything into a system or procure yet another tool. It's not that any one of these approaches is bad, but to scale the value of research and not just the number of researchers on the team, your research strategy should consider the interplay of all three: research outputs, people, and systems must be balanced and orchestrated to achieve strategic priorities (see Figure 2.1).

---

13 "Law of the Instrument," Wikipedia, Wikimedia Foundation, last edited April 8, 2024, https://en.wikipedia.org/wiki/Law_of_the_instrument

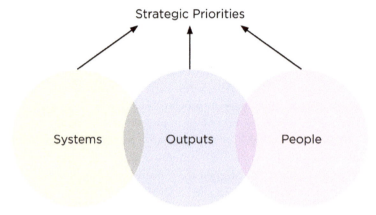

FIGURE 2.1
People add value, but outputs and systems must also be strategically set up to achieve priorities.

There's a tendency within the user research profession to focus on the *doing* of research as the sole avenue for increasing the value that research provides. More research equates to more value, right? Well, maybe…and it depends. But if hiring researchers is your only strategy, and if operation's sole focus is helping researchers work more comfortably or speedily, you'll also proportionally scale the cost of research to the business without necessarily increasing the value that it perceivably provides—or its valuation,[14] which is an important alternative to the more transactional word, *value*.

> **NOTE** IT IS ALL ABOUT VALUE
>
> *Value* is a word that's repeated countless times in this chapter (and book), and for good reason. Value can be created by delivering a successful product or service (superiority) or retained by spending less money on delivering it (efficiency), and often a simultaneous blend of both. The entire premise of a business, each department within it, and a strategy is to create more value. But in the world of research, whether you deliver value or not is largely to do with perception, or how people *evaluate* research. More under the heading, "Building a Foundation of Trust."

---

14 In a Rosenfeld Media video panel called "A Genuine Conversation About the Future of UX Research (Videoconference)," research manager Noam Segal mentioned the word *valuation* as an alternative angle to the conversation about research value, and he makes a great point. March 20, 2024, (https://rosenfeldmedia.com/sessions/a-genuine-conversation-about-the-future-of-ux-research-videoconference/

Research and the General Ledger

A good deal of the past decade's rapid growth of user research teams has been fueled by the following scenario on rinse and repeat: A leader from another discipline, say content strategy, interface design, or product management, would like to hire a researcher to work with them full-time. So, they acquire a head count and transfer it to the research department to facilitate the hire. In this way, one investment at a time, the research team grows from five to dozens or hundreds of researchers working across the organization. This is genuine growth, and it's not to be sniffed at, but there's a catch.

Although individual researchers will have (quietly) helped product teams to do better work, when executives are called to scrutinize operating costs, the research department will stick out like a sore thumb on the corporate general ledger: a complete record of the company's financial transactions, including the operating costs of individual departments. Transferred head count will tend to be reflected under the research department's operating budget, a total that can quickly add up to millions of dollars.[15] (If you run a research department, you should know this figure off hand.) At the level of executive management, a research practice can look incredibly expensive vis-à-vis the value delivered, or lack thereof. Of course, value has been delivered—*you* know that, and *I* know that—but it's locked up in the hearts and minds of the folks who were in direct contact with a researcher, or in reports scattered across tools and local drives. Also, the passing of time and too much going on means that stakeholders might forget that the great decisions they made were a result of meticulous research, not divine inspiration! "Besides, isn't research just chatting with customers, and anyone can do that. Right?" Research might be a cost center, but it can still be deemed to cost too much.

Smart research leaders are all too aware of this dynamic, so they make sure to align with corporate priorities and operate in ways that both distribute and amplify the value that research provides not just to more people, but over greater periods of time, i.e. longer than the length of a study (see Figure 2.2).

---

15 The other day, someone working at a well-known software company quoted that 5 million dollars is spent annually on research operating costs alone. The team has no research strategy and distributes its value.

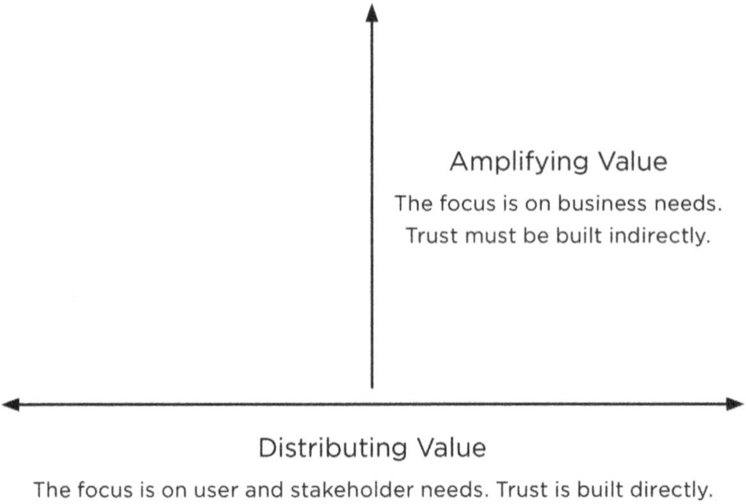

Amplifying Value
The focus is on business needs.
Trust must be built indirectly.

Distributing Value
The focus is on user and stakeholder needs. Trust is built directly.

FIGURE 2.2
You should balance efforts that distribute and amplify the value
of research.

Distributing Research Value

A *distributed model* means that researchers are distributed, in one way or another, across the organization. Researchers may partner with stakeholders who fund a researcher, or several, to join their team, or you might grow centrally to handle more research requests from across the organization. In both cases, the more stakeholders that you partner with and the more successful those partnerships, the bigger your research footprint will be. Either way, the value of research is typically felt by the people who directly benefit from the research, and by no one else, nowhere else, and at no other time. In other words, the impact tends to be discrete, local, and short-lived—and easy to forget in the bustle of teamwork. Also, trust is often founded on one-on-one relationships, which is valuable but hard to scale.

In the case of a distributed model, ops teams are often asked to support researchers by helping them to be more effective or efficient, which can be useful. But be aware that this approach on its own can make the research practice look even more expensive for little tangible value: Has helping researchers avoid the task of recruiting participants *really* delivered additional tangible value to the business? "Yes" is rarely an honest answer. Strategically, it's a good idea to prioritize adding value before seeking to do things better or save costs. A cheap cost center that doesn't add perceivable value is still a liability with little reward, so keep this in mind.

A distributed model is a genuine and valuable way to scale—it is even essential—but just as a million continuous sparks mean little unless they start a fire or are harnessed to drive an engine (that propels a vehicle in a particular direction to achieve a particular goal), so all of the research interactions and outputs produced as part of a distributed model should be harnessed to amplify the value that research perceivably provides.

Amplifying Research Value

Amplifications are usually operational efforts that gather up and curate existing value, or create new value, and deliver it en masse to a wider audience, say the entire company or a high-priority audience with limited attention and time, like executive management. An amplifying effort could be delivering a fully managed research library (see Chapter 6, "Long Live Research Knowledge"), hosting customer learning events, delivering cross-product strategic research reports, making observing live research as easy as pie, or empowering nonresearchers to skillfully spend time with customers. Amplifying research value could also involve delivering research efforts that directly align with corporate priorities, like the music streaming company's goal of expanding into five new markets: India, Japan, Brazil, Mexico, and Sweden.

Growing "vertical value" takes thinking laterally about the types of people that you hire, the things that they do, how the function of research and ResearchOps is structured, and how to shape research assets and experiences so that they deliver value to people who aren't directly involved with a researcher. In this case, because the research material is often consumed indirectly—there is no researcher to bond with—perceptions about research (and the research practice) must be proactively shaped so that research produced is also used.

When you devise a research strategy, consider what you might do—and not do—to both distribute and elevate research value across the organization and to top brass. Rather than resorting to hiring more researchers or doing more research without thought, or punting the value of individual research studies, aim to deliver scalable research operating systems that will reliably provide people with learning experiences en masse. In turn, these efforts will amplify the value of research both as a craft and a budgetary department. But, as already mentioned, for any of it to succeed, the perceptions, biases, and values about research—how to gain knowledge when needed, and who to trust with being right—must be set up for research to succeed.

Building a Foundation of Trust

Several years ago, while having lunch in a corporate canteen, a conversation caught my attention. Two product managers were lamenting that they couldn't work out how to apply the findings that were shared in quarterly quantitative research reports about their product. Though they admitted that the reports were meticulous—perhaps too meticulous—they felt that the insights were out-of-date and hard to action, and that researchers were out of touch. "They live in an ivory tower," they said. It broke my heart because I knew how much effort, smarts, and money went into producing those reports, and that the work was technically excellent.

If you consider the basic premise of this story, it's likely that you can recall at least one similar story of your own. It's unfortunately not uncommon for good research to lie fallow, fall on deaf ears, or for the investment to outweigh the benefits. Research can fail to deliver value for countless reasons, from working on the wrong thing or a mismatch in expectations and communication styles, to timing or a lack of trust or influence—never mind how smart or well-balanced your strategy and the resulting operating model.

Rather than talk about value, research professionals tend to use the word *impact* instead. In many ways, this makes sense. It implies that the final product or value of research is to impact or influence decision-making (for the better, of course). In other words, the value of research doesn't sit in the number of reports produced, or even in their veracity, it sits in the influence that the research practice can exert on the right parts of the organization, and eventually on the users' experience.

But as any politician or salesperson will tell you, to influence people you must first get them onside, which often means appealing to their existing values. This doesn't mean that you need to pander to skewed ideas around what constitutes good research, assuming your values and those of your stakeholders are not aligned. It does mean that, as part of your research strategy, you'll need to consider the gap between the influence you need and the influence you have, and how to bridge it. I've heard too many stories of research teams that have found themselves at odds with entire parts of an organization because ideas around what constitutes "good research" collide: "They don't understand what good research looks like." "They're doing it all wrong." "They have little respect for the craft." Whether the laments are right or wrong is beside the point. If the research practice is perceived to be snooty about research, too slow to collaborate, or a blocker of progress, it is unlikely that people will advocate for research or appreciate the value provided.

Playing to Existing Values

User research is all about understanding the needs of users, and then meeting them where they are. And it's exactly this approach that's needed to understand what consumers want, need, and assume about research. The intricacies of how you explore the existing value structures with important stakeholders is up to you—you likely already have access to excellent research skills. However you approach it, you should answer the following kinds of questions:

- **Why have people chosen to invest in research?** Organizations don't invest in things; people do—the CEO, your boss, a stake-holder—and they always have expectations. So, what are they? Remember that a strategy is specific, so your answer should be substantially more nuanced than "people want to learn about customers."

- **How successfully are you meeting current expectations?** To understand where things are going right or wrong, you will need to ask potentially vulnerable questions about performance, but you could also glean a great deal of information about engagement from operational metrics—say the number of research studies that a team invested in, or, if you have a research library, the number of research access requests within a specific period of time, and *who* made those requests. (See more in Chapter 10.)

- **Are their expectations realistic and well-informed?** As someone who works in the field of user research, you will know that while people often know what they want, they don't always know the best way to achieve their goals. It is your job to figure that out and then devise strategies for getting them on board.

Answering these questions can be tough and they can be highly political: researching the value of research means interrogating individual researchers' effectiveness and the effectiveness of the research leader, but it's better to face the music when you still have the time and capacity to change the tune! If this work isn't done sensitively and with buy-in from the person who owns the research, you may find yourself in hot water. For these reasons, it's best to hire a consultant to do this work and involve everyone in the process so that the benefits are made clear, and it feels like progress as opposed to an attack.

In a Nutshell

For the most part, the battle to create a demand for research is won. Yes, sometimes the demand is too much or too fast, or assumptions about how to meet the demand are misguided. But the fact that demand exists means that opportunity is right behind it. The challenge for research leaders now is to shape the demand, and this can only be done by taking a strategic approach to how a research practice delivers value to the business. So, you must:

- **Devise a research-specific strategy that outlines what the research practice will do—and *not* do—to deliver value that the business cares about.** Remember that strategy isn't just a noun (a document), it should also be a verb: a constant doing.

- **Let go of the notion that individual researchers or studies should be able to demonstrate an impact on the bottom line.** As a cost center, your broader valuation within the organization is vastly more important, so work on that first.

- **Relying solely on a distributed model to deliver value will result in the research practice looking extraordinarily expensive vis-à-vis the value delivered.** You can resolve this by leveraging centralized ops efforts that amplify the value of existing research assets and efforts, like customer events, live-research experiences, strategic research (as opposed to research strategy), or a well-managed library.

- **For any of these strategies to succeed, you'll need to invest in the culture of research.** The biases, perceptions, and expectations that drive what people want, and whether they trust and engage with the research that you deliver is vital. So, don't leave this to chance. (See Chapter 11, "Getting Priorities Straight.")

From Strategy to Operational

In November 2021, H&M, the second biggest fast-fashion company in the world, set a new climate target to reduce its supply chain emissions by 56% by 2030. The fashion industry produces 10% of all human-produced carbon emissions, "more emissions than all international flights and maritime shipping combined,"[1] according to the World Economic Forum (WEF), and it's the second-largest consumer of the world's water supply. So, H&M's move was important and industry leading. But setting a strategic goal, like halving emissions, is the easy bit. The real smarts lie in how well you can execute it, particularly when it involves changing how a global behemoth like H&M operates. One of the ways that H&M sought to achieve its strategic goal was to address a key environmental and business issue in the broader fashion industry and in its own supply chain: wastage.

"In total, up to 85% of textiles go into landfills each year. That's enough to fill the Sydney harbor annually," wrote the WEF. If you've not had the pleasure of visiting Sydney harbor—I live on its shores—it is *big*. While consumers throw a lot of clothing away, clothing retailers also regularly jettison excess stock. This is because they typically use a *push operating model*. In other words, they place bets on what the demand for a product might be, secure orders for it well in advance, and then they *push* their inventory toward consumers. Retailers hope that they'll strike the perfect balance between producing too much or too little stock, vis-à-vis demand, to minimize wastage and maximize profits. But demand forecasting isn't a hard science, and the levels of excess stock generated by retail fashion show that it leaves a lot to be desired.

To address wastage, H&M collaborated with the design and innovation company, IDEO, and together they mapped out the complex network of systems that make up H&M's global supply chain and the people who populate it, from designers to garment suppliers and logistics managers to people working on shop floors.[2] As a result, they created an algorithm that could more accurately predict demand and shorten the interval between sales and production, which no doubt required all sorts of other adjustments to how H&M operates—technology is rarely a solo solution. The pilot resulted in a 22%

---

1   Morgan McFall-Johnsen, "These Facts Show How Unsustainable the Fashion Industry Is," *World Economic Forum* (blog), January 31, 2020, www.weforum.org/agenda/2020/01/fashion-industry-carbon-unsustainable-environment-pollution/

2   Catharina Frankander, "Better Sales, Less Waste," *IDEO* (blog), www.ideo.com/works/h-and-m-kigumi

reduction in stock and a 34% increase in sales: it positively shifted H&M's environmental impact while improving the bottom line and freed up staff to add value in other ways. It was a win-win-win.

More than an education in the problem of fast fashion, H&M's story is a textbook perfect example of how to turn a strategic priority into an operational reality, whatever your context. The key takeaways are that a strategic priority should *always* lead to a shift in how an organization operates; without this crucial step, a strategy will fail to materialize and sustain. While it's tempting to think that big goals need equally big and shiny new initiatives, even relatively small adjustments to how an organization operates can lead to success. Also, a shift in one part of operations, like demand forecasting, can help achieve strategic priorities that might at first glance seem disconnected, like cutting supply chain emissions—making these sorts of connections is part of good strategy. Finally, a priority needn't be achieved in one grand sweep. Instead, a combination of multiple tweaks and transformations to operations might be just the thing that's needed to succeed.

FOOD FOR THOUGHT

A 1% CHANGE CAN MAKE A BIG SHIFT

To deliver the goals of a strategy, a business may need to do things that are bold and innovative, but goals are also often achieved by changing how a business operates day-to-day. This is a lot like author James Clear's notion of *continuous improvement*, which is "a dedication to making small changes and improvements every day, with the expectation that those small improvements will add up to something significant." In the world of operations, this approach can translate into adjusting or overhauling tasks that may seem mundane, but which are constant or have far-reaching consequences. Clear adds, "Improving by just 1 percent isn't notable (and sometimes it isn't even noticeable). But it can be just as meaningful, especially in the long run."[3] In fact, a 1% change made every day can make a 37.78% shift over the course of a year. The trick lies in choosing the right 1% shifts, or combination of shifts, to help achieve a strategy over time.

---

3   James Clear, "Continuous Improvement: How It Works and How to Master It," https://jamesclear.com/continuous-improvement

Strategy in Review

The first step in defining a strategy is outlining a limited list of goals to achieve, or strategic priorities. Then you must devise a strategy for how you'll achieve each of the priorities, as covered in Chapter 2, "Lost and Won on Strategy." In H&M's case, their goal was to reduce supply chain emissions, and minimizing wastage was one way to achieve it. To devise a strategy, you'll need to gain a deeper understanding of the problem—I'm preaching to the choir!—which will inform the kinds of 1% tweaks or outright transformations that should be made to how the organization operates, i.e., its *operating model* and the systems that innervate it. All of which you must track, constantly iterate, and communicate (see Figure 3.1). You'll learn more about operating models in, "Designing an Operating Model" later in this chapter.

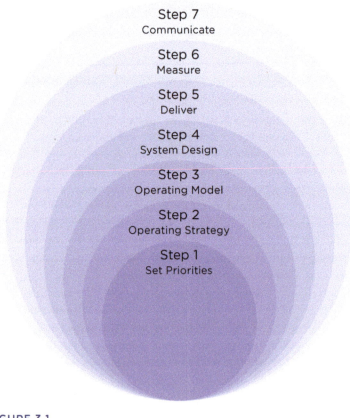

Step 7
Communicate

Step 6
Measure

Step 5
Deliver

Step 4
System Design

Step 3
Operating Model

Step 2
Operating Strategy

Step 1
Set Priorities

FIGURE 3.1

There are seven initial steps to turn a strategic priority into a way of operating.

While it's incumbent upon research leaders to decide where to invest, it's the role of the ResearchOps team to work hand-in-glove with leaders to put priorities into scalable action, which requires that ops folk should be skilled at more than admin and program management.

ResearchOps Is a Design Job, Too

It is often assumed that ResearchOps is an administrative role with little space for creativity and invention, but that couldn't be further from the truth. At its core, operations is a *systems design* role, and all good systems are specific, which implies strategy, too. Every system will require coordination to set up and administration to keep it running, like renewing contracts, keeping pay-as-you-go funds topped up, handling support, and more. But long before these tasks kick in, a system should be designed so that admin is minimal, it integrates with the wider research operating system, and it either supports a strategic goal or a standard business requirement like onboarding staff or keeping data safe.

I've met too many senior ResearchOps professionals who have never designed an operating system, but inventing, designing, sketching, and experimenting with how things will operate (by designing a research operating system) is an innate part of their job—ergo services and systems design's enormous contribution to ops. It should be second nature for operations folk to doodle systems to life.

So, the first person that you hire into a ResearchOps team should have systems design skills; they should *not* be an administrator, which is a common mistake. There are countless ResearchOps "teams of one," so if your systems-designer type will be the only person delivering ops, they will also need to administer the systems that they build, which can be a tough (but not impossible) balancing act. No matter how good they are at their job, the admin required to run the systems will eventually outpace their capacity to cover it all, so keep this in mind as operations mature. (Who to hire and when will be covered in Chapter 4, "Planning Realistic ResearchOps.")

None of this calls for an entirely new breed of professional, though. Apart from an in-depth knowledge of how research practices work—and *don't* work—all of the human-centered and systems design know-how and know-what required to make strategies operational have already been beautifully mapped out by the wider design profession. Once again, delivering ResearchOps requires adopting and adapting existing expertise rather than making things up from scratch, a regular theme in this book. And if you're already working as an ops professional and you can see that you have a gap, make services and systems design a focus for professional growth. (Note: I've chosen to use the words services and systems interchangeably. Services are, after all, human-centered systems, as are organizations.)

RECOMMENDED READING
SALUTATIONS TO SERVICE DESIGN

Service design is an essential ally for operations, with its focus on user-centered and end-to-end services and systems mapping and its invaluable library of templates, such as research journey maps and service blueprints. In the past, I've either done service design work myself, or, in richer times, I've hired service design consultants and with excellent results. If you don't have funds for the professionals, dive into a service design book, or better yet, sign up for a course. If you want to upgrade your skills or access templates **servicedesigntools.org** is a good place to start. You might pick up Lucy Kimbell's book, *The Service Innovation Handbook: Action-Oriented Creative Thinking Toolkit for Service Organizations*. IDEO U offers a course called "Human-Centered Systems Thinking," which looks ideal. **www.ideou.com/products/ human-centered-systems-thinking**

The Systemic Design Framework

One of the most influential design organizations in the world is the UK Design Council. The Design Council was inaugurated in 1944 by then British prime minster, Winston Churchill, to tackle economic recovery post World War II.[4] Since then, the Council has been the UK's national strategic advisor for design, and it has led globally in showing how good design can improve people's lives, drive the economy, bring people together, and enable businesses to do better business—or more impactful research.

4　"Our History," Design Council, www.designcouncil.org.uk/who-we-are/ our-history/

If you work in the field of human-centered design, you're likely to have come across the hugely influential *Double Diamond* (see Figure 3.2). It was popularized by the UK Design Council circa 2004, and it has been a hallmark of good design ever since. The Double Diamond[5] is a problem-solving framework that illustrates a four-phase nonlinear and iterative design process: discover, define, develop, and deliver.

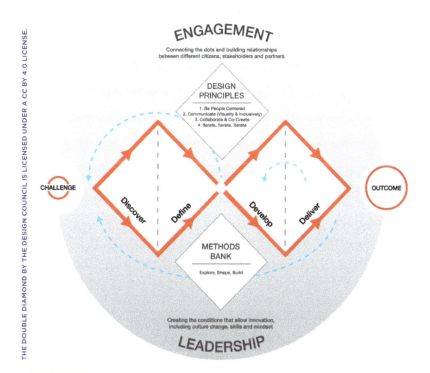

FIGURE 3.2
The Double Diamond visualizes the design process and outlines key principles, design methods, and the working culture needed to achieve enduring change.

In 2023, the Design Council's then Director of Design and Innovation, Richard Eisermann said: "The ascendance of fast-paced digital design, along with the complexities of the challenges designers are currently addressing with services and systems, have left the Double

---

5 "The Double Diamond," Design Council, www.designcouncil.org.uk/our-resources/the-double-diamond/

Diamond a bit short of breath."[6] So, the Design Council developed the Systemic Design Framework (see Figure 3.3), an even more marvelous match for describing the systems-oriented work of operations.

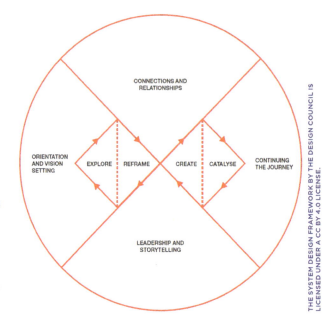

FIGURE 3.3
The Systemic Design Framework builds on the Double Diamond's iterative four-phase design process and recognizes the "invisible activities" that predict the success of complex systems.

Like the Double Diamond, the design process asks you to explore and reframe until a problem is well-defined, though the nomenclature is purposefully updated. What the Systemic Design Framework usefully adds is the "importance of the 'invisible activities' that sit around the design process: orientation and value setting, continuing the journey, collaboration and connection, and leadership and storytelling. These activities are rarely recognized, and therefore rarely resourced, and rarely done," wrote Cat Drew, Chief Design Officer at the Design Council. If you're an ops professional, you probably feel seen and heard! Also included in the framework, although not visually represented, are four characteristics that every systems designer should hone: systems thinker, leader and storyteller, designer and maker, and connector and convenor. It's a perfect list of the chief characteristics of a senior ops professional. Hirers, take note.

---

6   Richard Eisermann, "The Double Diamond Design Process—Still Fit for Purpose?," *Medium* (blog), May 11, 2023, https://medium.com/design-council/the-double-diamond-design-process-still-fit-for-purpose-fc619bbd2ad3

Whether you're designing a research operating system, an operating model, or a system or subsystem within it, the Systemic Design Framework is an invaluable ally. Print it out and put it in clear view because, when you're under pressure to deliver something without a clear purpose, as a silo, or in record time, it will remind you to consider the wider system and its parts, and the wider task at hand. (See Chapter 4.)

Designing an Operating Model

If you poll any number of folks working in business, you're likely to get as many variations on the definition of the term *operating model* as people who respond. One of the greatest challenges in writing this book was getting my head wrapped around the divergent views about what an operating model was versus an operating system, service model, plain old system, or service versus system, and more. In the end, outside of honoring traditions and not coming across as out of tune, I decided to keep things simple.

Typically, an *operating model* is defined as a high-level view of how an organization, like a research department or an entire business, will operate to create value. In line with that view, a 2019 *Deloitte Insights* article offered this definition, "Our definition of an operating model is simple but comprehensive: An operating model represents how value is created by an organization—and by whom within the organization."[7] But an organization is really just a system driven by a multitude of subsystems. So, in my view, an operating model is a plan for how a specific system, or set of systems, will operate and interoperate to deliver value. See Figure 3.4. For example, you might devise an operating model for how the entire research organization will operate, or how part of it will operate, say enabling embedded researchers to engage stakeholders more easily in research, or how part of *that* system will operate, like a research lab or insights library.

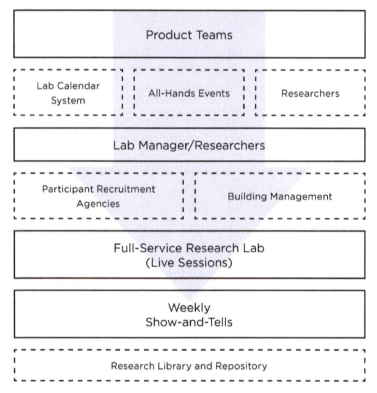

FIGURE 3.4

A fair amount of pedantry exists around the design of operating models, but they needn't be time-consuming or communicate every detail. This generalized example is intentionally simple and visualizes how a research team might operate a research lab and weekly show-and-tells to elevate their work.

7 Anne Kwan, Maximilian Schroeck, Jon Kawamura, "Architecting an Operating Model: A Platform for Accelerating Digital Transformation," *Deloitte Insights,* August 5, 2019, www2.deloitte.com/us/en/insights/focus/industry-4-0/reinvent-operating-model-digital-transformation.html

An operating model is usually represented as a visual schematic that gives a top-down view of the following factors:

- **Structure:** How the system will be divided into teams or units and how they'll interact.
- **Modes:** The modes or strategies by which the system, or parts of it, will operate, e.g., full-service versus self-service or push versus pull. (See "A Library of Operating Strategies.")
- **Processes:** What and how information, people, or tasks will flow or sequence.
- **Interactions:** Touchpoints that your system will have with other systems, actions, or people.
- **Protocols:** The decision-making framework, including policies, rules, and controls.
- **Tools:** The platforms and technologies required, and how they will interact and interoperate to achieve goals.
- **Measurement:** Where and how you will collect metrics, and what to do with them.
- **Mitigation:** Potential weak points in the system, and how to mitigate them.

Making this schematic will prompt you to consider not only how the overall system should operate, but also how the array of systems within it should operate. To design these systems, depending on how complex they are (simple systems might be quickly sketched), you'll need to use the sort of design approach laid out in the Systemic Design Framework along with service design resources and skills to map the front- and back-of-house operations, service touchpoints, and more.

More than just doodling schematics, when designing an operating model for a system, one of the most important decisions that you'll need to make is the strategies by which the system will operate to deliver value, save costs, and achieve goals. For instance, will the system operate on a full-service basis, or will it be self-service? Will it operate on a push model, as most fast-fashion retailers do, or on a pull model, say in response to a waitlist? The above *operating strategies* are widely used in the industrial, retail, and services worlds, and they're equally applicable to research.

A RESEARCH LAB BUILT BACK TO FRONT

In 2013, I was asked to build an in-house research lab for the UK's Government Digital Service (GDS). The research lab was instrumental to one of Leisa Reichelt's[8] strategic priorities, which was to make user research a team sport, i.e., everyone should be exposed to users taking part in research for at least two hours every six weeks.[9] One of the ways to achieve this was to provide a space in which people could easily gather to watch research live: a research lab, or even better, a studio.

Though the studio was a relatively simple system, even it had a conscientiously designed operating model that purposefully aligned with the strategic intent.

I did a Discovery phase and learned that most labs at the time featured a large room for the researcher and participant/s (to accommodate focus groups) and a small observation room. But the goal of GDS's lab was to get crowds of people through the door to watch one-on-one research, so I bucked the trend and used the larger room to house observers and the smaller room as the research space. As part of the Discovery, I visited a media company's swish research lab in London, and it featured a large observation room with all sorts of creature comforts, including a fully stocked bar fridge!

While I couldn't offer free sodas to government employees to showcase research and keep observer attendance high, I did install a *huge* television screen and excellent sound, and negotiated the budget required to hire a full-time A/V specialist who made sure that the observers' experience was excellent and managed the day-to-day tasks of running the lab. This meant that the lab was a *full-service* offering that had important touchpoints with the organization's existing calendaring system, staff events, and building maintenance.

---

8   Leisa Reichelt was the GDS's Head of User Research at the time and transformed how research was viewed and done, not just within the UK government, but globally.

9   Jared M. Spool, "Fast Path to a Great UX—Increased Exposure Hours," *Medium* (blog), October 5, 2015, https://jmspool.medium.com/fast-path-to-a-great-ux-increased-exposure-hours-afde796f2e43

A Library of Operating Strategies

When I first joined a band in 1996—I was a bass guitarist—the way that I listened to music changed. To this day, instead of hearing a song as a single sound, I hear it in layers: drums, bass, guitar, vocals, synths, and production tricks all fascinate and inspire me. Similarly, once you start thinking in systems, you'll notice how all sorts of systems layer and operate, from a hotel booking system to an ecommerce store, to the way that your local coffee shop takes orders.

An operating model can be made up of one or multiple strategies that together describe how an entire system will function. Just as musicians strike one note or riff off standard chord progressions, you should do the same using operating strategies. The following are the most common ones:

- Self-service versus full-service
- Standardized versus customized
- Volume versus variety
- Push versus pull
- On-demand versus request-and-response
- Centralized versus decentralized
- Automated versus manual

Self-Service Versus Full-Service

To provide people with access to your systems, you'll need to give them a way to engage. The service strategy that you choose to use is vital to the success of your system—it's a sink or swim decision—and the most critical selection of all is self-service versus full-service, or a hybrid of the two.

- **Self-service:** In the ResearchOps context, self-service means that your operations enable people to skillfully help themselves with things like participant recruitment—the prime self-service example. But self-service can also apply to using a lab, finding research reports, using a tool, sending participant thank-you gifts, or completing self-paced research training. The list is literally endless.

- **Full-service:** As it says on the can, a full-service model means that the stakeholder does little if anything to reap the rewards of the service: *you do it all.* In the context of research operations, you might offer full-service participant recruitment, research literature reviews to brief, or a research lab with a lab operator in-situ per the previously shared story.

- **Hybrid:** An operating model could also include both: a hybrid. You might offer a self-service option for most things, but a full-service option for needs that are high-priority or require specialist operations or skills, such as participant recruitment for accessibility research or to recruit participants who are VIPs or hard-to-reach.

Standardized Versus Customized

Standardization is the bedrock of scalable operations: in order to achieve any kind of efficiency and scalability, repeatable tasks must be standardized. To standardize means to define a standard set of practices, principles, or guidelines that define how something must be done. Recipes, wheels, plug points, charging cables, shoe sizes, and even algorithms and chat bots are examples of the millions of mundane things that are standardized.

Standardization is a function of scale that enables consistency, repeatability, modularization, automation, and task transferability—even transferring tasks to your users, as in the case of on-demand self-services. It's worth noting that standardized practices don't necessarily stand for best practice, but they should at least stand for good practice or good results in most cases, most of the time. The build/buy and standardize phase of operational maturity (see

Chapter 4) specifically prods you to build standardizations into your systems and make explicit choices about what parts, if any, should be customized.

Customization, on the other hand, means that something is produced to suit a particular set of needs at a particular time. It is usually a full-service experience, and it is custom fit to the context, so the results are consistently excellent. Whether something is standardized or customized (or *so* custom that it's bespoke) has a significant impact on the resources you'll need to drive your operations.

The tailoring world offers a nice illustration of the differences: when buying a suit, you can choose to buy a suit that is off-the-rack (standardized self-service), custom-made (standardized full-service), or bespoke (customized full-service).

- **Standardized self-service:** In this case, an off-the-rack suit is made to standardized sizes. You simply walk into the shop, try it on, and buy the best possible fit for your size.

- **Standardized full-service:** Here, you measure yourself at home and order your suit online. While you get a suit sewn to your exact size, the tailor simply adjusts a standardized pattern.

- **Customized full-service:** This requires a bespoke suiting service. The suit making involves specialized tailoring skills, hours of detailed work, and the luxury of a personal measuring and fitting service. The result is a perfect, unique, and effortless fit—and it's expensive.

In the context of ResearchOps, standardized self-service ops might include:

- Templates for research participant communications, such as a study invitation, confirmation, or thank-you emails, which may be automatically sent.

- A self-service research lab that operates on audio-visual presets and protocols, and has a well-understood etiquette.

- A standard operating procedure (SoP) or runbook[10] that can be used by an operations admin to self-manage an unfamiliar or irregular task.

---

10 Used widely in the IT management industry, a *runbook* (comprised of run sheets) is a documentation of routine procedures and operations that an administrator or operator must follow to maintain the network or IT infrastructure, or to complete a particular task, such as offering technical support.

Standardized full-service operations might include enabling:

- Researchers to submit a request to a participant recruiter who will manage the limited logistics of recruiting a standardized cohort of participants for them, say vegans in New York.

- A team to request that you deliver a standardized research training module at a particular team event or time—the training is standardized, but the experience is personal.

- Researchers to request an operations administrator to send one of three types of standardized thank-you gifts to a research participant, say an egift card, swag box, or coupon.

KNOW YOUR SUPPLY VERSUS DEMAND

It's a rookie error to offer a service before understanding, at least to some degree, your supply versus demand ratio—*all* the numbers—both current and forecast. I once hired a single participant recruiter to support full-service participant recruitment without first understanding how many people were doing research and, therefore, how many participants we'd need to generate. It didn't take long to realize that 500 people were doing research and were thrilled at the idea of participant recruitment that was easy and free, and delivered within two weeks. Of course they were! I now know that I would have needed a team of 50 participant recruiters to satiate the demand, and I'd hired just one. What an amateur!

To work out a supply versus demand ratio, fire up a spreadsheet and answer these fundamental questions:

- How many people will require the service?
- How often will they require it—and how many?
- How much of the thing will they need: money, seats, participants, administrators, thank-you gifts, field kits, etc.?
- Can you predict the demand and therefore the supply?
- Could existing data help you set a baseline for supply and demand, and, therefore, deliver sustainable operations?

Working this out will help you devise an operating model that's realistic to sustain and scale over time. For example, if the demand for something is high and the supply is low, you may decide to deliver a full-service system to tightly manage how the system (and supply) is used, or a self-service system that's centrally controlled and has limited access.

Customized full-service ResearchOps might include:

- Full-service participant recruitment delivered to a particular brief.
- Custom-designed participant thank-you gifts. Perhaps researchers can request that you send customized gift hampers to hard-to-reach or high-earning participants.
- Bespoke literature reviews in response to a unique brief.

Volume Versus Variety

The traditional operations world has a valuable dictum to follow: *low volume for high variety, high volume for low variety.* To use the suiting analogy one last time:

- The off-the-rack factory can produce standardized suits into the thousands (high volume for low variety).
- The custom online tailor can produce lots of suits (high volume), but not as many as the off-the-rack factory.
- The bespoke suiter may only produce a dozen or so highly customized suits per year (low volume for high variety).

FOOD FOR THOUGHT

VOLUME OVER VARIETY DRIVES ALDI

High volumes don't need to mean low quality. The German supermarket chain Aldi is a perfect example of a business that has chosen to deliver a low variety of products at high volumes, while retaining quality. "Stores carry less than 1,500 different items (a typical U.S. supermarket carries more than 25,000), and about 90% are branded under the chain's private labels. To offer extremely low prices and still earn a profit, the operations function must keep costs to a minimum, seeking every efficiency available while maintaining an assortment and quality of products that keep customers coming back."[11] Aldi is an example of operational mastery that's worth studying.

---

11 Marco Iansiti and Alain Serels, "Operations Management Reading: Operations Strategy," *Harvard Business Publishing*, June 27, 2013, https://hbsp.harvard.edu/product/8000-PDF-ENG

Push Versus Pull

You might recall that retailers, like H&M, do their best to forecast the demand for a product, secure orders for it in advance, and then push their inventory onto customers to sell as much of their inventory as possible. This is a *push system*; its opposite is a *pull system*.

- A *push system* as mentioned earlier, produces products based on forecasts of future demand. In ResearchOps terms, you could forecast the types and quantity of research participants that might be needed in the next quarter or year, deliver an engagement program to attract those types of participants, and then push those participants toward people who are doing research. You could also use a push system to manage an inventory of swag purchased to use as participant thank-you gifts, otherwise known as *incentives*.

- A *pull system* responds to an existing demand. Many ResearchOps teams manage recruitment operations on a pull system. This means that researchers, or operations if recruitment is full-service, seek to pull the right research participants from a variety of recruitment pathways when needed.

Well-rounded participant recruitment operations should ideally use both push-and-pull systems to deliver enough participants to meet run-of-the-mill recruitment briefs and those that are more variable. (See Chapter 5.)

On-Demand Versus Request-and-Response

You could choose to offer a service experience that is either *on-demand* or *request-and-response*.

- *On-demand* means that someone is being delivered a service or asset immediately upon request—a highly efficient way of working if you can get it right. On-demand services are often driven by self-service operations and are highly standardized and automated. You might enable researchers to download a self-paced training module, which, once complete, gives them instant and automated access to a particular tool. You might standardize the operations needed to recruit a generic cohort of participants—say those vegans who live in New York—and then enable researchers to recruit on-demand.

- *Request-and-response* indicates that someone must submit a request, often via a service or support desk, and then wait for a response. Ideally, they'll receive a response within an indicated response time, also called a *service-level agreement* (SLA). Request-and-response is the doorway to whatever kind of service you want: the full concierge experience or prodding someone in the right direction so they can help themselves. You might offer researchers the opportunity to request access to a research tool or securely held data, or to request a custom recruit or a literature review.

Centralized Versus Decentralized

Centralized operations mean that systems are set up so that ResearchOps directly controls and delivers them. In the case of tooling access, for instance, ResearchOps will retain total control of who gets access to a tool and how they get access.

The alternative is that you *decentralize* the work of delivering services to partners within the organization or externally. Decentralizing is different from outsourcing a one-off job: hiring an agency to recruit for an ad-hoc participant recruitment brief isn't the same as decentralizing participant recruitment wholesale. Instead, decentralizing is about creating a standardized relationship that delivers ongoing efficiencies. Decentralizing works best when you're transferring a regular and repeatable administrative task that's well-documented to another person or team. You can also decentralize more variable tasks so long as they don't require extensive internal knowledge. You might:

- Outsource administrating access to research tools to whoever does this task for the wider organization. Usually, it's IT or workplace technology who, to take on the job, may require a standardized list of tasks documented in a runbook or SoP.

- Outsource participant recruitment *en masse* to an external agency. Again, decentralizing is not about hiring an agency to handle ad hoc jobs; it's about ongoing and repeatable efficiencies.

- Empower a select group of researchers, or people who do research, to support their teammates with research or ResearchOps questions. This is a kind of decentralized support.

- Decentralize the ongoing marketing initiatives needed to grow an in-house participant recruitment panel to your organization's marketing team or an external marketing agency.

Decentralized operations can lift weight off your operations, but it's not a free ticket. You'll still need to manage contractual and financial relationships, and you'll lose valuable hands-on insight and control of your operations. To stay well-informed, pay close attention to your operating metrics and foster strong and constant relationships with your service users and partners.

Automated Versus Manual

Automation relies heavily on standardization: even artificial intelligence is a string of complex standardizations or algorithms. If automation doesn't frustrate people in their moment of need, say conjuring up memories of an annoying robotic phone assistant, automation can eliminate repetitive tasks that are inefficient and uninspiring to administrate, bring inhuman speed to your operations, and enable the ability to scale exponentially. As great as automation is, reserve it for tasks where a human touch or brain won't be inordinately missed, and where it's not a do-or-die situation.

If you choose to include automations in your workflow, make sure that you understand how the automation works both technically and experientially so that it doesn't break or damage your workflow or technology. Finally, it can be easy to "set and forget" automations, but they do need to be maintained, so keep a running audit of the automations that are live across your workflows and check them regularly.

At this point in history, automations are best used to augment a service rather than replace human beings—even the AI sensation, ChatGPT, agrees! (see Figure 3.5).[12] There are however countless opportunities to automate for greater efficiency. For starters, you could automate:

- An email to remind research training attendees that they're signed up to take part in a training session, thereby reducing no-shows.

- A Slack message to let stakeholders know that a research session has been scheduled, and they should attend it, thereby increasing observer numbers and research engagement.

- The transfer of data from one place to another via an API[13] to maintain the up-to-dateness of a participant recruitment panel or augment it with additional ethically sourced data.

---

12 "Introducing ChatGPT," OpenAI, November 30, 2022, https://openai.com/index/chatgpt/

13 API stands *for application programming interface*, which is software code that enables two applications to share information.

FIGURE 3.5
For something tongue-in-cheek: even ChatGPT doesn't think that it can eliminate the job of a researcher!

But what about *manual* operations? Automation shouldn't get all the excitement. While manual operations typically feel old-school and don't scale well, manual operations can be the basis of a success story when a human touch is either necessary or noticeable. For instance, if you're running a customer advisory board (CAB) that includes high-touch and hard-to-access VIP customers, you'd do well to keep communications highly personal, which means customized and manual.

The Interrelatedness of It All

By now, you've likely noticed that the operating strategies are linked. For instance, if demand is high, you'll need to work at high volumes, unless you've got significant resources, or the ongoing need is bespoke. To satiate a high-volume demand, you'll need to standardize to a limited selection of offerings (low variety) that are delivered self-service and accessed on-demand or via a request-and-response system. You might choose to automate all or part of the system, and high-volume products are often delivered via a push model.

Participant thank-you gifts offer a nice example of this. If you're supporting thank-you gifts at a high volume, you'd do well to support a small selection of standardized offerings. Egift cards, charity donations, and one type of standardized swag box should take care of most needs. If your volumes are large enough, this model will allow you to take advantage of the economies of scale with a limited selection of vendors, or just one vendor, if you can swing it.

Making Invisible Results Visible

One of the trickiest aspects of scaling research operations, is showing the value delivered behind the scenes. Making invisible results visible isn't impossible, but it is an art. For example, IDEO's article about their work with H&M featured two easy-to-understand numbers at the top of the screen (see Figure 3.6): a 22% reduction in stock during a pilot program, and a 34% increase in sales. In simpler terms: we saved money; we earned money (or added value).

PROGRESS

22 PERCENT

Reduction In Stock During A Pilot Program

34 PERCENT

Increase In Sales During A Pilot Program

FIGURE 3.6
Simple and tangible; this IDEO example of storytelling using numbers hits the nail on the head.

You don't have to know anything about H&M, the fashion industry, or good business management to know that those numbers are good. They are simple, tangible, and meaningful without much context, and they are an excellent example of how to share impact that would otherwise be behind-the-scenes and, therefore, invisible. You'll learn more about operating metrics in Chapter 10, "Money and Metrics."

In a Nutshell

In his book *The Creative Act: A Way of Being*, Rick Rubin says "The more we pay attention, the more we begin to realize that all the work we ever do is a collaboration."[14] He continues to write that "It's a collaboration with the world you're living in. With the experiences you've had. With the tools you use. With the audience. And with who you are today." Though Rubin is famous for producing music, his words are relevant to the creative work of turning strategic priorities into systems that change how a research practice operates, which is the only way to achieve change, and every strategic priority requires change. These are the points to remember:

- **Strategic priorities don't always need big and bold new initiatives to succeed.** Consider shifts and tweaks to existing operations, or a combination of efforts—big and small, new and old—to achieve goals.

- **Take a human-centered and systems-thinking approach to designing operations.** This involves divergent thinking (generating a wide range of ideas) and convergent thinking (narrowing down to the best ideas) at each iterative stage.

- *Everything* **is a system of sorts, and every system should have a considered strategy for how it will operate to deliver value.** This requires defining how value will be exchanged, whether via a self- or full-service, push-or-pull, etc., model.

- **The best operational shifts are the ones that deliver results with the least amount of expense, effort, and disruption.** But the better you are at achieving this goal, the better you'll need to be at measuring and communicating achievements. Per the Systemic Design Framework, it takes leadership and storytelling skills to make excellent ops visible.

---

14 Rick Rubin, *The Creative Act: A Way of Being* (New York: Penguin Press, 2023), 89.

Planning Realistic ResearchOps

H ere's a big number: 6.67408 m/s². That's the *Big G* or the *gravitational constant*[1] as first defined by Isaac Newton in his Law of Universal Gravitation formulated in 1680.[2] The fact that gravity exists on planet Earth won't come as a surprise, but when you look up into the night sky and see the moon and various planets, like Mars and Venus, which are viewable with the naked eye, and even the sun during daylight hours, it can be easy to forget that these celestial bodies, and countless more, are being governed by the fundamental force of gravity. At the heart of it all is the sun, a colossal star that dominates the solar system. It holds the planets in orbit, and they, in turn, exert their own influence on the sun and neighboring planets so that everything is held on track, as if by a giant pulley system.

The strength of a planet's gravitational influence is determined both by the masses of relating planets and objects and the distance between them, per Newton's law. For instance, Jupiter, the largest planet in our solar system—it is twice the size of all other planets combined—has a gravitational pull so strong that it draws at least 64 moons into its orbit, and it can tear asteroids apart. Even celestial objects that are light years away play a role in shaping the orbits of other planets and objects, like meteors and stars, albeit diminished with distance.

The solar system may seem abstract to ResearchOps, but when it comes to operating systems, it is the biggest of them all, so it too holds valuable lessons. To deliver operations that scale, every system, subsystem, end user, and stakeholder, and how they all interact (plus their mass and pull) should be considered at the outset (see Figure 4.1).

At first glance, that might seem an overwhelming task, but it is a complexity that's resolved with a simple conceptual framework for delivering operations that are both complete and realistic to sustain—longevity being a prerequisite for scaling. ResearchOps is rarely simple or straightforward, which is part of its professional allure, but it's a lot easier to handle when you are armed with an elemental chart and a map of the universe. This chapter may have started with outer space, but its focus is on bringing big ideas, and complex research operating systems, down to earth.

---

1 Gravitational constant, denoted by *G*, is a fundamental constant used to quantify the strength of the gravitational force between two objects.

2 Keith Cooper, "What Is the Gravitational Constant?," *Space.com*, last updated September 22, 2022, www.space.com/what-is-the-gravitational-constant

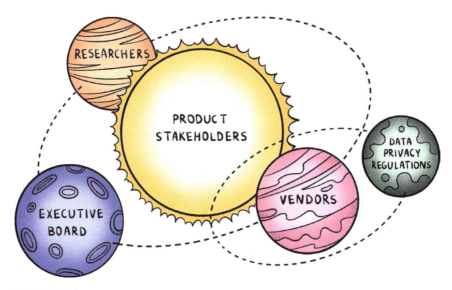

FIGURE 4.1

A ResearchOps initiative can only succeed if it functions in unison with the universe around it: users, stakeholders, vendors, and outside forces like legislation and technological advancements.

The Eight Elements of ResearchOps

To keep track of the elements, scientists use the periodic table (see Figure 4.2),[3] a chart that shows the elements that all matter is made of, like argon, krypton, and boron—or for something more familiar, sodium or iron. There are 118 known elements represented on the periodic table, and it helps scientists to quickly find basic information about an element just by looking at it. It is *really* smart. Relative to organic chemistry, ResearchOps is vastly less complex, but it still makes things so much easier to handle when they're broken down into distinct parts—or elements.

---

3   "Periodic Table of Elements," National Center for Biotechnology Information, https://pubchem.ncbi.nlm.nih.gov/periodic-table/

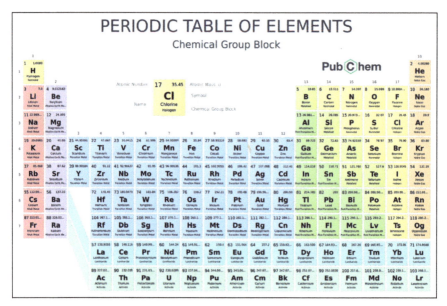

FIGURE 4.2

The periodic table of elements is used to understand the relationships between elements, predict their properties, and guide scientific experiments.

For a fundamental understanding of research operations, think of it as having eight elements, all of them essential to the functioning of a scalable research practice, all critical to the formation of a rich research culture. The following are the eight elements:

- **Participant recruitment:** Accessing people to take part in research is every researcher's biggest pain point, and it's often the inspiration for bringing operations onboard in the first place. By giving researchers more efficient, effective, and respectful access to the right participants, operations can contribute enormously to the speed and efficiency of research.

- **Knowledge management:** Research knowledge (or insights, by another name) is the key product and defining asset of research. It's vital that operations learn how to, or hire people to, master organizational knowledge to scale the value that research provides.

- **Onboarding and support:** To keep everything running and everyone humming, you'll need to support people who use your systems, tools, and services from day one. Training, playbooks, service desks, support channels, office hours, and onboarding are all on the research support smörgåsbord.

- **Tools and vendors:** From sticky notes and Sharpie pens to field kits, labs, agencies, participant recruitment platforms, and tools for data asset management (DAM), these days there's a tool for every part of the research workflow. It's the work of research operations to make sure that the right research tools and vendors are available, compliant, useful, and used.

- **Ethics and privacy:** Researchers collect a lot of personally identifiable information (PII), which means that ethics, legal compliance, and privacy are pervasive in ResearchOps. Here, the focus is on enabling research practices that both do no harm and manage the organizational risk involved in managing people's data whatever your scale.

- **Money and metrics:** Money is power, and there is mastery in getting it and spending it. Operations should deliver financial administration, but it should also deliver financial energy, savvy, and excellent fiscal partnerships. Money is also metrics, and monitoring how money is spent can unearth compelling narratives about research in the organization and your operations—a power that you should wield to your advantage.

- **Program management:** Most research organizations are at the pointy end of can't-do-enough-research-fast-enough, but ResearchOps can put systems and services in place that help researchers and the wider research organization set priorities straight. Equally important to enabling excellent execution is executing the right thing.

- **People and skills:** The chief operational asset of every research organization is researchers, and if the doing of research is democratized, people who do research—*and*, of course, operations specialists are equally important, too. To deliver reliable and impactful research, you'll need to attract, onboard, grow, motivate (and keep motivating) top talent, whatever their job title.

If you've been watching the ResearchOps space for a while, you would have seen similar lists in the past. It is not uncommon to hear people aspire to hiring a person to cover each element, or worse, one person to cover it all, but there's a catch. A nice tidy list gives the impression that the eight elements of ResearchOps are siloed things, but that's not the reality. Just as few things in the universe are made of just one element—water or H_2O is comprised of one oxygen and two hydrogen atoms—so few things in ResearchOps are made up of just one element, too. Instead, each of the elements is closely or distantly connected one to the other, and often dependent and interdependent on one another.

Whatever you do in ResearchOps, whether you are building an entire operating system or procuring a single tool, you will need to simultaneously address every one of these elements to a greater or lesser degree. This intrinsic interoperability is best illustrated as a Venn diagram, and it's critical because it represents how you should plan, partner, and hire to deliver realistic research operations.

RESEARCHOPS FRAMEWORKS

When it comes to ResearchOps, there are two (related) frameworks that do well in sharing a general lay of the land. The first was created by the ResearchOps Community, which I founded in early 2018. By the close of that year, we had launched and completed the #WhatisResearchOps initiative, which inspired and coordinated 60 researchers who ran 34 workshops around the world to ask user researchers a key question: What is ResearchOps? The data that emerged from those workshops (along with a survey) informed the #WhatisResearchOps framework, which I published on *Medium* in late 2018.[4] That framework has been used by researchers and leaders around the world to gain operations buy-in and to kick-start their work. Since then, my framing of ResearchOps has evolved and matured to better match the reality of building scaled-up operations, but the original framework still has utility.

The second framework comes from researcher Emma Boulton who was keenly involved in the #WhatisResearchOps project. Her eight pillars of user research[5] differ somewhat from the eight elements presented in this book, but they enhanced the premise of the #WhatisResearchOps framework beautifully.

I stepped away from leading the ResearchOps Community in 2019, leaving it to others to grow and lead.[6] A serial maker of communities, since 2019, I have run a members' club that's dedicated to connecting full-time ResearchOps professionals globally. Unconventionally, it's called the Cha Cha Club: chacha.club.

---

4   Kate Towsey, "A Framework for #WhatisResearchOps," *Medium* (blog), October 24, 2018, www.medium.com/researchops-community/a-framework-for-whatisresearchops-e862315ab70d

5   Emma Boulton, "The Eight Pillars of User Research," *Medium* (blog), July 11, 2019, www.medium.com/researchops-community/the-eight-pillars-of-user-research-1bcd2820d75a

6   Kate Towsey, "I'm Stepping Aside as Leader of the ResearchOps Community: Thoughts, Why and What Next," *Medium* (blog), February 15, 2019, www.medium.com/researchops-community/im-stepping-away-as-leader-of-the-researchops-community-75d1ad268673

The Venn Diagram of ResearchOps

A Zen master visiting New York City goes up to a hot dog vendor and says, "Make me one with everything." The hot dog vendor fixes a hot dog and hands it to the Zen master, who pays with a $20 bill. The vendor puts the bill in the cash box and closes it. "Excuse me, but where's my change?" asks the Zen master. The vendor responds, "Change must come from within."[7]

It's a fun joke, but it's also a useful introduction to the Venn Diagram of ResearchOps (see Figure 4.3).

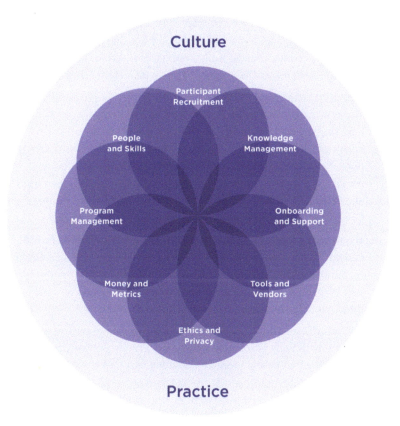

FIGURE 4.3
The Venn Diagram of ResearchOps illustrates the codependence of the eight elements.

---

7   "Some Zen Jokes," thezensite, www.thezensite.com/ZenEssays/Miscellaneous/ Zen_Jokes.html

Not unlike the *mandalas*, a circular figure that represents the universe in Hindu and Buddhist symbolism, the Venn diagram represents the universe of research operations in its most complete and accurate form. The Venn diagram offers a cognitive nudge to remember that if you work on one element, you must work on them all—and perhaps that lasting change comes from how you operate, per Chapter 3, "From Strategy to Operational."

All Is One: Delivering a Recruitment Panel

Say you want to provide researchers with an in-house *participant recruitment panel* to meet a strategic priority, like retaining streaming customers if you were Disney or improving brand loyalty if you were Uber. Although you may start with participant recruitment, soon enough you'll be working on tools and vendors, ethics and privacy, onboarding and support, and so on. Here's how delivering a participant recruitment panel breaks down:

- **Participant recruitment:** There are a raft of tasks involved in building a panel, including defining a panel strategy—good systems are always built to achieve specific goals—and putting operations in place to attract the right people to sign up and remain interested via an engagement strategy. You'll learn more about building an in-house panel (as one recruitment strategy) in Chapter 5, "People to Take Part in Research."

- **Knowledge management:** When building a panel, you should consider how your participant recruitment taxonomy, i.e., name, location, job title, and more, relates to your research repository and library taxonomies, assuming that you're at that level of maturity when it comes to knowledge management.

- **Onboarding and support:** You'll need to develop pathways and materials that signpost newcomers toward the recruitment panel (awareness), onboard them so that they can make use of it when needed (enablement), and support them in using it (help). This trifecta applies to both researchers, administrators, and panelists.

- **Tools and vendors:** Unless you have access to an existing tool, perhaps a purpose-built tool or a CRM, you'll need to select and procure a panel management platform. This will involve ongoing work with legal compliance, data security, financial management, and onboarding and support. You might also build templates (a type of tool) into the service to increase efficiency and continuity for everybody involved, participants included.

- **Ethics and privacy:** *So* much data is collected in research that it is rare not to have privacy partners involved at some point. In this case, they will need to approve how data is stored, used, and maintained both within the panel and across the workflows that enable it all. You may also need to develop a panel-specific consent form or two.

- **Money and metrics:** Participant recruitment is rarely free, so you'll need to constantly manage the money needed to fund it all. You'll also need to set up systems to monitor metrics like usage, response-rates, opt-outs, and new signs-ups to track and communicate how the panel performs and to aid the financial forecasts necessary to support budget planning.

- **Program management:** Providing access to the right participants at the right time is the goal. One way of doing this is to make sure that the right kind of participants, and related operations, are available to meet the needs of upcoming programs of work. Panel metrics will help you understand the demographics of your participant population and where you've got gaps vis-à-vis participant requirements.

- **People and skills:** You'll need to help people use the panel proficiently by providing ways for them to upskill. You'll need to offer how-to training, i.e., how to use the tool, but you may also need to offer craft-oriented training on topics like targeting participants or writing successful recruitment screeners. You might also need to hire specialists to maintain the panel.

It's common to view the eight elements of ResearchOps, or similar frameworks, as discrete areas to work on, but that's a mistake. Just as you can't scale research—only research systems scale, per Chapter 1—you can't deliver participant recruitment, knowledge management, or ethics and privacy, for instance. They are nouns, not verbs. (If that seems pedantic, it's also a vital cognitive shift.) Though you might focus on knowledge management, everything you do should be geared to meet a particular operational need—a *doing* that adds value. To address knowledge management, you might build a library, a repository, or a community of practice, all of which will involve all the elements to a greater or lesser degree. (Note: A librarian is an excellent person to hire if you're building a library but less useful if your goal is a community of practice, depending on the librarian, of course.) Managing research knowledge will be covered in Chapter 6, "Long Live Research Knowledge."

Failing to see and plan for elemental interconnectedness, and mistaking elements as anything more than ingredients, means that programs of work are often *wildly* underestimated (and underspecified) resulting in those who are delivering it feeling overwhelmed, those who are investing in it feeling underwhelmed, and those who are hired feeling misplaced. Once you see each element of ResearchOps as a lens—a universe that holds degrees of every other element within it—and you tie initiatives to strategic intents rather than generalized frameworks for thinking, you'll be able to scope operations realistically, deliver value that perceivably meets business needs, and hire the right people into your team.

FOOD FOR THOUGHT

MYTH BUSTING: A RESEARCHOPS TEAM OF ONE

If ResearchOps can live by any dictum, it's this: there's no such thing as a ResearchOps team of one. Even a one-person "team" will need a rich network of internal and external collaborators from across specialties, like finance, legal, education, marketing, and information science, to deliver their goals. One person *can* lead many, but it's unrealistic to think that one person can be a financial whizz, a librarian, a legal nerd, and a data, marketing, and community-building expert (all participant recruitment skills). These are specialist areas that require degrees, or at least significant experience, not to mention oodles of time.

The Venn diagram of ResearchOps describes the eight specialist areas you'll need to simultaneously consider to deliver a research operating system or discrete efforts. The eight elements define the ingredients, but nothing exists independent of *time*—another big theme in physics. While agile delivery (Discovery, Alpha, Beta, Live, Retirement) offers an invaluable framework for planning and delivering systems of all kinds, its generalist nature means that it doesn't provide a view of how operating systems develop. The *three phases of operational maturity* fill that gap.

The Three Phases of Operational Maturity

One day, I was catching a bus in Los Angeles and took a second too long to give the fare to the driver. His reaction was instantaneous, "This is a *fast* city! A *fast* town! You gotta move *fast-fast-fast!*" he shouted. And that was in the 1990s. Every book, article, or blog post on culture and business makes mention of the increase in pressure, speed, and competition these days. It's a dog-eat-dog world, they say, and it's no different in research and research operations. The pressure is on to deliver things cheaply and effectively, and preferably *today*. Reliable and scalable operations don't manifest in days, weeks, or even months. However, by delivering well-chosen transformations that are tied to a value-added operational need or a strategic priority, it is possible to deliver operations that delight, or at least suffice, in the here-and-now and aren't a compromise months or years down the line. Getting this right starts with strategy, tactics, and designing a good operating model, per Chapter 3, but you will not see results if you don't execute realistically.

To support realistic planning, it's useful to understand that every system should progress through *three phases of operational maturity* (see Figure 4.4). The three phases are:

- Build/buy and standardize.
- Specialize and optimize.
- Keep the lights on (KTLO).

In combination with the Venn diagram of ResearchOps, an understanding of the phases will help you:

- Realistically scope operations initiatives.
- Illustrate the full investment required.
- Understand whom to hire to see you through all or part of a journey.
- Plan for the long-term maintenance and support required to sustain and develop mature operations, a step that is too often forgotten.
- Spot opportunities for partnership, growth, innovation, and efficiencies.

FIGURE 4.4
The three phases of operational maturity aren't linear. Instead, once operations are in place, they should follow a continuous and circular sequence of specializing, optimizing, and measuring, while you keep the lights on.

> **NOTE** **MEASURE IN EVERY PHASE**
>
> Measurement isn't explicit in the naming of the three phases but consider it implicit: measuring your progress is important in *every* phase of operational maturity. The moment you build or buy something, consider how you'll measure its input, throughput, output, and experience quality. You'll learn more about measuring operations in Chapter 10, "Money and Metrics."

Build/Buy and Standardize

Eric Ries's bestselling book *The Lean Startup* says, "You don't have to be in a garage to be in a startup. The concept of entrepreneurship includes anyone who works within my definition of a startup: a human institution designed to create new products and services under conditions of extreme uncertainty."[8] If you're a ResearchOps professional, that definition should sound all too familiar. Hardly a month will go by without being asked to deliver something brand-spanking new, which is an important (and fun) part of the delivery

---

8   Eric Ries, *The Lean Startup* (New York: Crown Business, 2011), 28.

cycle. Just don't misconstrue delivery as done. In this phase, you're looking to turn your operating model into something real, so you will need to build or buy the following:

- A tangible thing, or set of things, to bring your operating model to life.
- Standardized processes that enable the system to deliver value that's repeatable and scalable.
- Systems for collecting data so that you can track and communicate the return on investment.
- Methods for engaging with the users, stakeholders, vendors, and partners so that you don't miss a beat.

Build/buy and standardize is the start-up phase of ResearchOps; it's driven by the speedy delivery of shiny new things, and it occurs when creativity, growth, and impact seem most obvious and exciting. Think fast-paced experimentation, learning, creativity, and bold choices, even controlled chaos. Much like the ethos of lean startups—you could think of this as *lean ops*—you'll want to learn, prototype, and run pilots, trials, or early-adopter programs (without being wasteful) just to discover what works best. But don't read this as being trigger happy: your work in this phase will dictate how you operate in the long term, so it should be led by a research or ResearchOps strategy, and ideally both.

Depending on your intent, by the end of this phase you might have built or bought something like the following:

- A research participant panel or a lab.
- Infrastructure that enables product managers to regularly spend time with customers.
- A research tool to support diary studies along with the training, support, and compliance needed to help it succeed.
- Playbooks or how-to guides, i.e., standardized processes and templates, for how to do research well.
- Communities, guilds, or councils to help shape research standards.
- Documented standard operating procedures (SoPs) to enable efficiency, continuity, task transferability, and change resilience across the research operating systems.
- Agreed ways of working to support productive relationships with stakeholders, partners, and vendors.

- The capabilities required to measure operational usage, even if rudimentary to start.
- And if you do all the above, a research operations team to manage it all.

> **NOTE** **BUYING ALWAYS INVOLVES BUILDING**
>
> Buying things, rather than building them, tends to work better when you're experimenting because the internal investment is minimal. But remember that buying always means building things into the internal infrastructure—like plumbing a sink into a kitchen—so consider this work when you plan, too.

Build/buy and standardize involves a lot of on-the-ground work and managing down because you're forming and storming[9] your operations, your team, and your partnerships. But you'll also need to manage up: You're selling ideas, informing strategies (see Figure 4.5), and angling for buy-in from leaders. Unfortunately, many organizations that are new to ResearchOps hire junior administrators or project managers to fulfill their first operations role, which couldn't be more misguided. In this phase, you'll want strategic change makers, systems designers, innovators, and senior program managers: visionary leaders who dive in, get stuff done, and aren't scared of the unknown.

FIGURE 4.5
The build/buy and standardize phase must be driven by a strategic priority.

---

9 "Tuckman (Forming, Norming, Storming, Performing)," Wageningen University and Research, www.mspguide.org/tool/tuckman-forming-norming-storming-performing

BUILDING/BUYING AND STANDARDIZING RESEARCH TRAINING

The set of scenarios shared in this chapter are based on a skills-development initiative that I delivered while leading the ResearchOps team at Atlassian. It was part of a much wider strategy that succeeded in getting hundreds of Atlassians involved in understanding customers both in collaboration with researchers and on their own.

Say you work in a fast-paced organization that hasn't got enough researchers to meet the demand. *That* old lemon! It's a scenario that likely isn't hard to imagine. As a result, other disciplines have taken on the role of doing research and, while they're doing their best, you're aware that the quality isn't always up to par. You decide to offer research training to up-level their skills and form bonds while your research capability grows. You're not sure how the training will be received or the effort versus value, so you run a pilot with designers.

As part of the pilot, you hire a training vendor and develop a *standard operating procedure (SoP)* and assets so that the logistics involved in delivering the training, like scheduling, marketing, and recording attendance, are efficient and repeatable. Once the pilot is complete, you review attendance metrics, cost per participant, and feedback on how the training was received. You look for anecdotal evidence as to whether it was impactful in improving research quality or not, noting that the full value might only be seen in the long term, and take note of the state of relationships.

The information you gathered in this phase has helped you learn what worked and what didn't, whether the effort versus the value is worthwhile, and the extent and type of demand. It's also helped you consider the best long-term operating model: You're questioning whether you should offer training modules as self-paced videos (self-service), live group sessions, one-on-one coaching (bespoke), or a mix of all three (hybrid).

As a result of the pilot, you understand how to improve your SoP, the skills and resources you'll need to scale the training in the long term, and the demand you should expect to meet. You're also now equipped with tangible stories, metrics, and feedback to help you get buy-in to continue or pivot the service, or to shut the experiment down if it simply didn't work.

The build/buy and standardize phase is often where operations efforts get stuck. Teams tend to build or buy things but fail to plan how they'll be maintained and scaled in the future. As a result, well-intentioned and even well-executed efforts can slowly fall apart over time—this is especially true in the case of knowledge management. To deliver operational lift that lasts, plan for the future, consider current and future supply and demand, look for standardizations in everything you build or buy, and set up mechanisms to measure your operations so that you know that your effort is worthwhile from day one.

Included in this chapter is a matrix that will help you do all of this with a birds-eye view. (See "The ResearchOps Planning Matrix©.")

Specialize and Optimize

So, you've built/bought and standardized a system, and it is delivering value, or you've inherited something from someone else and you'd like to refine it, which is a good idea. The second phase of operational maturity is about specializing (tuning or reusing systems to meet more discrete and valuable priorities) or optimizing systems (finding additional efficiencies), or a mix of both. In other words, it's about homing in on delivering maximum business value. In this phase, you will want to:

- Further standardize, automate, decentralize, or transfer tasks to create more efficiency.
- Create the capacity to deliver operations at higher levels of quality and specificity.
- Optimize partnerships and form long-lasting bonds that are mutually rewarding.
- Deliver measurable impact and success stories to help gain buy-in for new priorities.
- Consolidate systems to take advantage of minimization and the economies of scale.
- Watch for trends in your metrics: a year or more into measuring, you'll have enough data to *inform strategy*, which is an important milestone.
- Simplify decision-making so that you're spread less thin.
- Buy back capacity, which you can use to build or buy new things or scale existing operations thereby delivering even more value.

Where the build/buy and standardize phase is driven by strategy, the ongoing metrics and feedback generated in this phase should be used to shape strategy (see Figure 4.6). This is important, so read this twice: if you enabled the ability to measure your operations in the build/buy and standardize phase, which you always should, the data that your operations generate—who, when, where, how many, and how much—will provide opportunities to add quantified value to the conversations and decisions of senior leadership, particularly around research strategy. For the first time, operations can become an *informant*: a source of valuable operational information about the research organization and the research in the organization. As a result, operations can be seen as more than just the person or team that delivers vendors, tools, and handbooks, or, even worse, assists people to do research, which isn't the goal of operations at all. It is a critically important and triumphant moment for any operations leader, so don't miss the opportunity.

FIGURE 4.6

The constancy of metrics available in the specialize and optimize and KTLO phases offers the opportunity for ResearchOps to inform the research strategy and other stakeholder strategies. This is a pivotal moment, so don't miss it.

The specialize and optimize phase is characterized by patient observation, analysis of feedback and metrics, and smart refinements. In this case, the risk-loving change makers so valuable in the build/buy and standardize phase will likely get bored, so you'll want to put detail-oriented specialists in charge. In the case of research training, you may hire a senior education specialist to design education programs, manage ongoing progress, and make key strategic calls. I did this and would do it again.

SPECIALIZING AND OPTIMIZING RESEARCH TRAINING

The research training sessions that you delivered were well-attended and impactful. Designers are now more thoughtful in how they approach research, and the training has helped them connect with the research organization, which has strengthened collaboration and buy-in at all levels. But there's a concern. While 70% of designers attended one or more courses, attendance is on a steady decline. You realize that while the courses were valuable, you've saturated the existing audience, so you decide to optimize training for design graduates and new hires. You don't want to settle only on designers attending the courses, however, so you adjust your marketing and networking to attract product managers to attend the training, another important cohort.

Feedback suggests that the training was overwhelming—it covered too many topics—so you look for ways to specialize. You decide to home in on research for design evaluation and user interview techniques, which will be useful for designers and product managers, respectively. By specializing and optimizing the training and limiting the number of courses you're running, you now have time to evolve the training. You can create new partnerships to ensure that courses are included in graduate programs and new-hire onboarding, and you can spend time meeting with product management stakeholders to form a connection.

You're also now in conversation with the organization's learning and development team (L&D) about transferring the admin tasks involved in delivering training to them, or at least replicating their operations. This would not just be efficient, but it would also consolidate the research up-skilling experience for all employees.

The metrics, feedback, and conversations that the training continues to generate are valuable in helping you and leadership make fruitful and strategic connections across the organization. The metrics are also proving invaluable in showing the impact that the research practice is making on customer understanding in general.

Keep the Lights On

Every time you flick a switch to turn on a light, multiple things must happen for the darkness to disappear: there shouldn't be a blackout in your area, the electrical bill must be paid, the electrics in your house must be in good order, and the bulb itself must be working. Even a lightbulb requires a series of constantly maintained systems to do its job well. Keep the lights on (KTLO) is as it says on the box: it is the phase of mastering the ongoing administration needed to keep everything running. Nothing is ever done-and-dusted in ResearchOps, and it is vital to make sure that the routine work of KTLO doesn't result in complacency, a state that will ultimately absorb operational efficiency and kill value. The goal is to settle but not slumber.

At first glance, KTLO seems the least exciting of the three phases, but it offers a continued and growing opportunity to inform decision-making and, if you've maximized efficiency, the capacity to explore new goals (audacious goals, even!) and innovations. In this phase, you're likely to:

- Build an administrative arm to your research operations team—*finally*.

- Produce consolidated quarterly and annual operating reports to interrogate and communicate both research and research operations' impact.

- Spot opportunities for efficiency and the former's mortal enemy, creeping inefficiencies. Operating reports, along with an ear to the ground, will help you do this regularly.

- Manage up even more boldly; forge even stronger alliances that enable even greater efficiency and buy-in, as well as growth opportunities.

- Tackle one or two big hairy audacious goals (BHAG[10]), the kind of goals that you're only able to consider because of your reputation and maturity.

- Keep a sharp eye on innovations in industry: invest in technologies and projects that are experimental, which will see you return to the build/buy and standardize phase once more.

---

10 The term *BHAG* was coined by James Collins and Jerry Porras in their book *Built to Last: Successful Habits of Visionary Companies*, published in 1994.

In this phase, you'll want to build a diverse operations team to sustain and mature your operations: your team will include designers and makers, managers, coordinators, and administrators. To maintain and mature operations, you'll want to hire dependable people who enjoy known quantities, checklists, and manageable workloads, but who also have an eye for growth and opportunity. In this phase, your SoPs and runbooks will come into their own, enabling administrative continuity when someone is out on leave or when someone leaves entirely (task transferability is key to scalability and continuity).

It's worth remembering that KTLO doesn't mean that you should stagnate and stop iterating and innovating. More often than not, you'll find that once you deliver the things you've built or bought, the optimization and KTLO phases will run simultaneously, and your maturing operations will spawn new ideas that will involve building or buying new things. It's a flywheel of continuous maintaining, maturing, optimization, and innovation. This is peak maturity and fun when it comes to delivering research operations.

The three phases of operational maturity and the Venn diagram of ResearchOps will help you plan operations both vertically (across elements) and horizontally (across time.) More than a static framework, this type of thinking should become innate to how you operate—how you plan, innovate, respond, partner, hire, and make decisions every day. But there's one last step to add: the foundations for successful operations are best secured in the planning phase, which requires that you compact a multidimensional journey that can take months, and often years, into a one-page plan.

As a result of your efforts, the research training operations are now specialized, efficient, transferable, and scalable. Your most important stakeholders have experienced first-hand the impact of the training program, and you're able to show your impact via regular operating metrics and trainee feedback. Getting buy-in to hire an education specialist to run the training program in the long-term is now comparatively easy, and you'll soon also hire someone to administer training logistics such as course scheduling and attendance—the bits that haven't been adopted by L&D.

Your efficiency means that you're now able to consider new training opportunities, and your metrics and intel are providing interesting ideas, like creating "snackable" training content that can be shared via your organization's social channels. This may attract a segment of time-poor audiences to attend the full training programs and gradually upskill them while they're in the thick of things.

Also, instead of exclusively teaching people how to do original research, you might teach them how to do a quick literature review of research that already exists in your library or help people decide when to work with a bona fide researcher rather than DIY.

The ResearchOps Planning Matrix

The ResearchOps Planning Matrix (see Figure 4.7) combines the eight elements of ResearchOps and the three phases of operational maturity to give you a 10,000-foot view of what you'll need to do to deliver operations that are sustainable and, therefore, scalable. Ideally, you'll work through the planning matrix when you are in the Discovery phase.

Problem Statement or System Name:	Build or Buy
	What do you need to build or buy to set up operations?
Participant Recruitment	
Knowledge Management	
Onboarding and Support	
Tools and Vendors	
Ethics and Privacy	
Money and Metrics	
Program Management	
People and Skills	
What resources will you need?	

FIGURE 4.7

The ResearchOps Planning Matrix will give you a 10,000-foot view of what you should think about to plan operations that are realistic and complete.

Standardize

What could you standardize to make operations more efficient or effective? Could anything be automated, documented, turned into a template, or constrained to set criteria?

Specialize and Optimize

Does anything already exist, i.e., tools, partnerships, services, etc., that you could adapt to meet this need? If something already exists, could it be optimized to meet the need even more effectively?

Keep the Lights On (KTLO)

What will you need to do daily, weekly, monthly, and/or annually to "keep the lights on" and mature the system over time.

As with many tools, the matrix can be used in multiple ways: at the start of an initiative, to jog creativity and planning should you hit a roadblock, to think through a small initiative or a complex system, to remind you of the lay of your ResearchOps land, or to help you navigate change. Just as it's difficult to teach someone how to fly an airplane or shoot a hoop using words on a page alone—these are skills that require experience, nuance, and practice—so it is tricky to teach you how to use the matrix in a paragraph alone. Even so, the matrix isn't difficult to pick up and use if you commit to experiment and play and think that there are no rules. And there aren't.

NOTE **A RESEARCHOPS PLANNING MASTERCLASS**
I advise companies and run a masterclass on planning ResearchOps systems using the matrix, but, if that's not available to you, experiment with it because it's sure to help you. **katetowsey.com/masterclasses**

In a Nutshell

As famed British statistician, George Box once said, "All models are wrong, but some are useful."[11] While models and frameworks are indispensable in making calculated leaps into new actualities, they rarely represent the full dynamism and complexity of reality. And yet models offer a sturdy scaffolding on which to build your own custom-fit world—they eliminate the blank page and illuminate a reliable direction. Whether you are diving into ResearchOps for the first time or you're a seasoned pro, use the models shared in this chapter not just as gizmos for planning systems but as archetypes for how to deliver scalable research systems that aren't just efficient, but also influence the value that research perceivably provides. The models are the following:

- **The eight elements of ResearchOps.** These are participant recruitment, knowledge management, onboarding and support, tools and vendors, ethics and privacy, money and metrics, program management, and people and skills.

- **Venn diagram of ResearchOps.** It's convenient to view ResearchOps as a series of individual things—pillars, elements, systems, things to do—but it's not the reality. A list is indispensable, for sure, but your efforts will fall short if you assume that it represents reality. "Unus pro omnibus, omnes pro uno" is the ideal ResearchOps motto; it's Latin for "One for all, all for one."

- **Three phases of operational maturity.** The most obvious phase of ResearchOps is the phase in which new things are built or bought. But if you don't plan ahead to specialize and optimize operating systems and KTLO, you will find it hard to sustain (and scale) research in the long term.

continues

---

11 Wikipedia contributors, "All Models Are Wrong," *Wikipedia*, The Free Encyclopedia, last edited February 7, 2023, https://en.wikipedia.org/w/index.php?title=All_models_are_wrong&oldid=1220273024

In a Nutshell (continued)

- **ResearchOps Planning Matrix.** It's useful to pull the eight elements and the three phases of operational maturity into a unified framework for planning operations. With a bit of practice, the matrix can be used to design large research ops systems of all shapes and sizes. Even if it just gets you thinking differently, it has done its job.

The first four chapters of this book focus on overarching concepts like scaling, strategy, systems, and models for how to think operationally. While this is invaluable information—you can't scale research or deliver ResearchOps without it—you'll also need to know how to put individual elements into action. The following eight chapters focus on each of the eight elements of ResearchOps, one chapter for each, and they're more tactical in nature. For instance, how *do* you deliver knowledge management efforts, like a research library, that delivers value and stands the test of time? What should you do to fix participant recruitment? (Is "fixing it" even the right approach?) And what *must* you do to make ethical and legally compliant research a part of everyday life? It's all in this book, so read on.

People to Take Part in Research

Say you work in a fast-growing company that has dozens of researchers working across multiple products and scores of employees who are being encouraged to spend time with customers. Leadership has asked you to "fix participant recruitment," a request that's often the impetus for hiring a ResearchOps expert in the first place. To address the need, you procure a list of recruitment agencies and tools, and deliver adjacent assets like a scheduling tool, a consent form, training, and a handbook. Templates, SoPs, and automations are set up, so that recruitment is as easy to do and administer as possible; you even tick all the boxes with the data privacy team! The resources are delivered to hundreds of employees via a self-service operating strategy. You know that setting up a

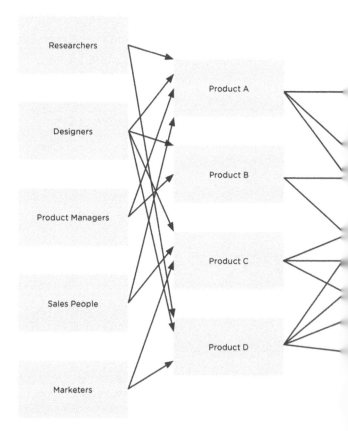

FIGURE 5.1
Most companies will have a lengthy list of variables, which impacts not only the number but also the variety of recruitment briefs. The longer the list of variables, the greater the number of potential combinations.

full-service recruitment desk is out of the question because you don't have the capacity to support or scale that kind of effort. Finally, communications are carefully crafted, and effort is put into support and change management. Life should be good—it is a textbook way to scale, after all. But, despite all of your efforts, it seems that your messages, emails, and conversations are littered with complaints that recruiting participants is still too difficult. I have a confession to make: This was me, circa 2020.

Scaling participant recruitment *is* hard, because it's a dichotomy. To deliver scalable operations of any kind, you *will* need to standardize things, and yet participant recruitment is riddled with variety (see Figure 5.1)—there is little standard about it.

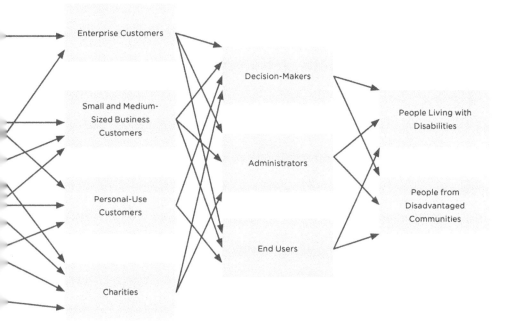

Whatever your context (B2C or B2B[1]), you'll need to provide access to an array of participants, from dime-a-dozen users to VIPs and people living with disabilities. Each of these cohorts will require a unique mode of engagement, which will impact the infrastructure, style, and flow of logistics so that each is unique—or nuanced, at least. If research is democratized, your operations must cater to a variety of disciplines each with different expectations as to what "research" means, never mind "participant recruitment." Finally, though researchers might typically do remote usability testing or user interviews, a variety of methodologies like longitudinal studies, field research, and unmoderated research are also on the table, each requiring alternate logistics.

If your head is spinning, it should be. Delivering recruitment operations that scale and shift with the business strategy, meet ever-changing needs, and foster respectful interactions is no small feat, but it is a possibility that is within reach. In addition to leveraging all-singing, all-dancing technologies, you'll need to:

- Standardize a million unique recruitment journeys.
- Make finding the right people as easy as possible.
- Address the ten steps of participant recruitment.
- Deliver logistics that are inclusive.

FOOD FOR THOUGHT

SCALING HUMAN EXPERIENCES

Henry Ford, the founder of the Ford Motor Company, is quoted as saying, "Mass production is craftsmanship with the drudgery taken out of it."[2] Likewise, it is the job of operations to take the drudgery out of mass participant recruitment, not the craftsmanship. To deliver strong research, researchers must be intentional about who they engage to take part in research and, though well-orchestrated operations may iron out friction, you should

---

1 It's typically easier to set up participant recruitment operations in B2C (business-to-consumer) contexts because consumers are easier to access—depending on the product, of course. B2B (business-to-business) contexts provide a unique set of challenges and opportunities, such as using company-owned data to access people who interact with the business in different ways: as decision-makers, administrators, or end users.

2 "Popular Research Topics: Henry Ford Quotations," The Henry Ford, www.thehenryford.org/collections-and-research/digital-resources/popular-topics/henry-ford-quotes/

not seek to eliminate researchers' personal investment with your over-the-top assistance. Instead, your operations should empower researchers to recruit on their own, just without the hassle of constant unknowns and administrative bottlenecks.

Standardize a Million Unique Journeys

As William Edwards Deming, the late American engineer, professor, and thought leader in the realm of quality and management, so aptly said, "Standardization does not mean that we all wear the same color and weave of cloth, eat standard sandwiches, or live in standard rooms with standard furnishings."[3] He continued, "Homes of infinite variety of design are built with a few types of bricks, and with lumber of standard sizes, and with water and heating pipes and fittings of standard dimensions." So, the trick to delivering scalable participant recruitment is this: you must find out what your bricks, lumber, and sandwich loaves are; then you must help people to use them.

To define the basic building blocks of a participant recruitment strategy, you must use the following steps:

1. **Define the most common participant cohorts.** Who are the broad populations that researchers most commonly need to engage? How broad can you go before the cohort becomes too generic to be useful? Think of this list as your *participant taxonomy*. Indeed, you may use your knowledge management taxonomy to inform this work and vice versa—and taxonomical skills to do it. Your most common cohorts will also likely match that most controversial of research assets, research personas. If you have an accurate set available, put them to good use. If you don't, look for funding to do that work.

2. **Make sure to include high-value cohorts.** You may come across cohorts that aren't commonly recruited but are high value—say big spenders, executive customers, or celebrities. Consider whether you want to provide a unique recruitment workflow for these cohorts simply because the rewards are high, or the risk of getting things wrong is significant.

---

3 John Hunter, "Standardization Doesn't Stamp Out Creativity," *The W. Edwards Deming Institute* (blog), October 5, 2015, https://deming.org/ standardization-doesnt-stamp-out-creativity/

3. **Prioritize cohorts.** The most important step of any strategy is deciding what to do and what *not* to do (see Chapter 2, "Lost and Won on Strategy"). Here, you must define a short-as-possible list of high-priority cohorts for which to deliver specifically designed recruitment workflows. Note: Common and high-priority cohorts alone may not accurately define where you should put energy. You must equally consider adjacent factors like organizational priorities, particularly those defined in the research strategy, and investment versus the reward.

4. **Consider the implications of methodology.** What are the most commonly employed research methodologies? Say qualitative and quantitative, or moderated and unmoderated research? How does the methodology impact, or cause variation to, how high-priority cohorts should be sourced and engaged throughout the recruitment process?

5. **Understand the needs of various disciplines.** It's a mistake to assume that all disciplines have the same expectations or needs. In my experience, product managers are often less interested in *research* participant recruitment and more interested in what could be called "customer speed dating."[4] Before jumping to conclusions, seek to understand the needs of the people who will use your recruitment capabilities.

6. **Decide on an operating model for each cohort.** Should you take a self-service approach, or would a full-service or hybrid model be more appropriate? There *are* instances where full-service recruitment delivered in-house is more effective than utilizing an agency or a platform, particularly if recruitment involves accessing customers who are hard-to-reach or have high-touch needs. But consider your approach carefully.

7. **Design the logistical workflow.** What practices and protocols can you put in place to support recruiting a particular cohort? What are the unique needs of that cohort? You'll learn more about the ins and outs of sourcing cohorts under the heading "Make Finding the Right People Easy," and you'll learn more

---

4 To generalize, researchers tend to want to recruit highly specified cohorts to take part in well-planned studies. Designers tend to want to routinely recruit similar cohorts of users to take part in usability testing. Product managers tend to want to regularly chat with a person, and not a neatly defined cohort, to keep up a discovery routine or to investigate a hot theme.

about standardizing recruitment logistics under, "Address the Ten Steps of Participant Recruitment."

8. **Look for operational similarities across workflows.** What logistical assets could you standardize to address the collective needs of multiple logistical workflows? Perhaps egift cards and charity donations would satisfy most thank-you gift needs. Perhaps you can produce a single consent form that's suitable to use in most research instances.

9. **Note where standards won't look after important needs.** There will be points at which a standardized asset simply won't cut it. If the cohort is common or important enough to warrant special attention, include it in your strategy and planning to make provisions.

If you follow the above nine steps, you will have shaped a solid participant recruitment strategy. But the success of a strategy lies in how well it's executed, and when it comes to participant recruitment, execution is *all* about how easily and reliably researchers can find people.

In its fifth annual State of User Research report,[5] User Interviews said that "The vast majority of researchers in our audience (97%) experience some type of recruiting pain. The most common challenge is finding enough participants who match their criteria (70%), followed by slow recruitment times (45%) and too many no-shows/unreliable participants (41%)." By defining a list of common cohorts and standards, you'll have taken a big step in the right direction. But giving researchers a list of resources called *customer panel, User Research International,*[6] or *Customer Advisory Board* won't help them navigate toward the right people.

Make Finding the Right People Easy

If you've spent time in a big city, it's likely that you've visited an area that was notable for a particular culture or craft, say Japantown in San Francisco for sushi and karaoke, the Garment District in New York, which is famous for its fabric stores, or De Wallen in Amsterdam, one of its best-known red-light districts. Even as a stranger to a city, it doesn't take much to get a sense of the kind of people,

---

5   Katryna Balboni, Morgan Mullen, Olivia Whitworth, and Holly Holden, "The State of User Research 2023," User Interviews, www.userinterviews.com/state-of-user-research-2023-report

6   User Research International, www.uriux.com

food, and activities you're likely to find if you were you to visit these places, and a map or travel planner will give you a good sense of the best route to get there (see Figure 5.2).[7] Conversely, if you know what you want to do ahead of time—Chinese for dinner?—the thematic nature of these areas will help you know where to go, even if you don't have an exact restaurant in mind. In essence, these places help connect you with the right people because they are placed on a map, and they're branded.

FIGURE 5.2

Even with limited knowledge of San Francisco, if you wanted fish and chips for lunch, it would make sense to try the Fisherman's Wharf first.

7   Wikimedia Commons contributors, "File: San Francisco districts map.png," Wikimedia Commons, https://commons.wikimedia.org/w/index.php?title=File:San_Francisco_districts_map.png&oldid=799184720 (accessed March 19, 2024).

When it comes to recruitment, it's common to forget our urban capabilities in lieu of resources and tools—ergo Chapter 1's ("Research Does *Not* Scale—Systems Do") focus on civic design. Suddenly finding people is all about resources and tools. But a panel labelled *customers*, however well-organized, tends not to communicate *who* a researcher might find there. To empower researchers to self-navigate toward the right people, you must select the most effective ways that researchers can access people. Then provide researchers with a mental or physical map or a decision tree (see Figure 5.3 on the next page) to help them understand where particular cohorts "live" and how to respectfully reach them.

A key part of a recruitment strategy is choosing the resources that you'll use to provide researchers with access to common cohorts. You want to be minimal, but not mean. These are the most common tactics, and each has a unique set of pros and cons:

- Agencies
- Participant recruitment platforms
- In-house panels
- Built-in panels
- Pop-up requests or intercepts
- Social recruiting
- Recruiting colleagues

Each recruitment tactic can be uniquely operated to meet vastly different needs, so think carefully about which tactic/s you choose to access which people. An in-house panel could be standardized to help hundreds of researchers access everyday users via generic communications, but it could also be geared toward accessing people who are typically hard to reach using highly personalized communications. The options are literally endless, so be mindful about your audience, i.e., researchers *and* participants, and what you want to achieve.

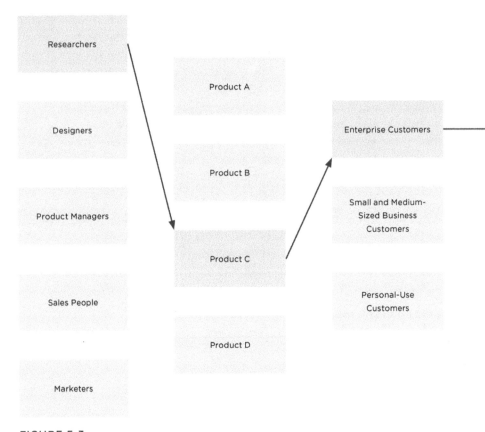

FIGURE 5.3

A decision tree can help researchers navigate toward the populations they need to meet, understand the best protocols to engage participants respectfully, and work well with partners.

Agencies

Agencies are often the first port of call when providing researchers with access to participants, and for good reason. Agencies offer a full-service recruitment experience without putting pressure on either researchers or operations, and they can usually source from a diverse range of participants. Perhaps you work for a research agency or a consultancy that needs access to music lovers one week, mothers-to-be another, and car enthusiasts the next. In this case, agencies are an excellent option. Agencies are also useful if you need to recruit participants who are high-touch, time-poor,

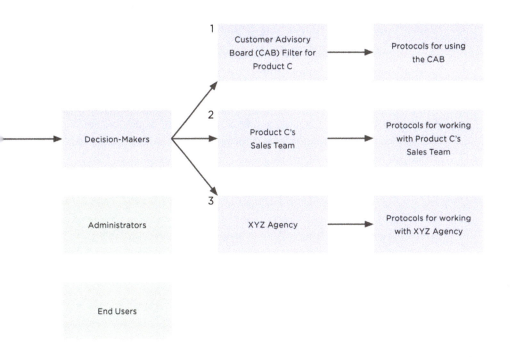

or hard-to-reach, say business executives, heart surgeons, or high-rolling gamblers. Also, specialist agencies may be able to access vulnerable people, like children or people with difficult life stories. In just about every context, it's worth having an agency or two on your books, but not all agencies are made the same, so do some research and choose the partners who will best meet your needs.[8]

---

8 In his book, *The Field Study Handbook*, Jan Chipchase offers a useful checklist for screening participant recruitment agencies (see page 76).

Participant Recruitment Platforms

The new breed of participant recruitment platforms[9] are a huge win for both researchers and ResearchOps. Platforms provide a user-friendly interface for researchers to traverse the qualitative recruitment workflow (see "Address the Ten Steps of Participant Recruitment") without leaving the tool, and they're a godsend for self-service recruitment. Outside of providing access to a commercial panel, platforms often offer the ability to host in-house customer panels within that same user interface, which means for double the firepower via just one vendor—a *massive* win for operations.

As wonderful as these platforms are, they do come with a couple of caveats. If the commercial panel's demographics aren't suitable to your researchers' needs, screener response rates may be too low to warrant researchers' efforts. Make sure to check that the panel can provide access to common cohorts before onboarding it and wasting researchers' time (and yours). If you do decide to onboard the panel, give researchers upfront guidance as to its strengths and weaknesses.

> **NOTE** **PARTICIPANT QUALITY IS EVERYONE'S CHALLENGE**
>
> Participant quality is a constant challenge for commercial panels, and, if too many participants aren't who they say they are, or no-shows are prevalent, word can quickly spread that a platform's gone sour—a notion that tends to stick. This can be tough to handle, but it's worth noting that *all* recruitment platforms battle the same challenges (even those that are in-house), and it's rarely effective or efficient to bounce from vendor to vendor. Instead, choose your vendor carefully by asking them for details of their panel demographics, quality control measures, and average response rates, and then commit to work closely with them to iron out any problems.

---

9   The most well-known and established recruitment platforms are User Interviews, Respondent, Askable, Great Question, and Rally. The rise of AI means that platforms are emerging that promise to cut out the "hassle" of finding people—living and breathing human beings—at all, like Synthetic Users. Call me a luddite, but the name says it all.

In-House Panels

In-house customer panels are often seen as a panacea to participant recruitment, and the rise of souped-up recruitment platforms mean that building them is now pretty easy. But even with the aid of these fantastic technologies, maintaining the growth and quality of a panel still takes significant effort. After you consolidate the costs of staffing, marketing, thank-you gifts, and platform fees, the budgetary benefits of an in-house panel can dwindle quickly. So, choose to invest in panels because they are the most effective or only recruitment strategy, *not* to save money. If you work in an organization that has a broadly consistent set of common cohorts, an existing community like a fan club, and a strong brand presence like Figma for designers, Peloton for fitness enthusiasts, or Squarespace for website developers, an in-house panel can be a fantastic utility. In this case, deliver well-labeled panels that match the participant cohorts outlined in your participant recruitment strategy. Even with filtering capabilities, a blob-like list called "customers" is generally less successful.

Built-in Panels

Many research tools like UserTesting, Maze, or Optimal Workshop offer built-in recruitment panels as part of their tool, which makes a lot of sense—a built-in panel gives people doing research a seamless research workflow in one interface. From an ops point of view, built-in panels have a nice benefit: almost every research organization will need a usability testing tool, and, if their panel is useful, ops can provide researchers with easy access to a pool of people without adding operational overhead. Particularly in the case of unmoderated research, *professional participants*—people who make a living, even if on the side, by taking part in research—are a common feature of built-in panels. It doesn't take much to find popular videos on YouTube that explain how to game user testing panels to earn more money. The notion of recruitment that enables honest partnerships with real people can go to rot in this case. Still, if researchers are made aware of these caveats and have other options available, and if recruitment criteria are generic, built-in panels have their place. Finally, don't feel pressured to purchase access to a built-in panel simply because you're procuring the tool. Most tools have a BYO option, too.

Pop-up Requests or Intercepts

The internet is filled with *pop-up requests* or *intercepts* to rate your experience, sign up to a newsletter, make use of a discount code, and, on occasion, take part in research (see Figure 5.4). A fin-tech researcher might want to interview people who have just opened a new bank account, so they set up an intercept to pop up at that exact point in the customer's journey. Intercepting someone, whether digitally or in person, can be an effective method for recruiting cohorts based on recent activity, and it's generally faster and cheaper than other methods. While intercepts have a lot to offer, if they add noise to an already noisy experience, disrupt customers at a critical point in their journey, or have a style or tone that is asynchronous with the context, they'll fail to recruit participants and even chase them away entirely. But used well and for a pointed purpose, they're an excellent recruitment opportunity. (Warning: Don't use intercepts to build a general recruitment population; it's a misuse of its specificity and not a good reason to interrupt a journey.)

> **Would you be interested in participating in a remote research session?**
>
> It will take around 5 minutes to complete from start to finish. It will help The Home Store design the future of your online experience.
>
> By clicking 'Yes' you agree to the research Terms & Conditions.
>
> **Yes**   No

FIGURE 5.4
A good recruitment intercept outlines the basics of the opportunity (a 5-minute remote research session) and, for good measure, confirms the name of the company (The Home Store).

It's likely that you'll need to partner with other teams, like marketing or product or web asset owners, to make sure that the intercepts are appropriately formatted, branded, and used. To help bring them onboard, open the conversation with a plan that shows respect for the brand and acknowledges the need to control noise, and then grow the relationship from there. If you decide to support intercepts operationally, you should provide researchers with the following assets:

- A standardized tool to create and manage intercepts, such as Intercom.[10] If pop-ups or intercepts are already used within the organization, you might save yourself effort and forge new partnerships by using the same tool.

- Templates that help researchers set up intercepts in ways that are brand consistent.

- Guidelines for setting up successful intercepts, including a workflow for data management and approvals, if needed.

- A centrally managed protocol for governing intercept frequency and quantity across the product space so that the user experience isn't too interrupted and noisy—let's be honest, intercepts always interrupt.

Social Recruiting

Sometimes, researchers will need to use multiple avenues to access the participants they need, and recruiting socially can be a valuable option. Depending on the social channel that researchers choose, they might access quite different cohorts (think TikTok users versus LinkedIn), so consider the channels that you want to support. Social routes include:

- Social media, like X (formerly Twitter), LinkedIn, or Craigslist

- Fan clubs, communities, or forums

- Account managers and salespeople who might be willing to connect researchers with customers

- Internal marketing lists

- People on the street, or *guerilla research*—although that term seems to have gone out of favor

---

10 Intercom, www.intercom.com

Because social recruiting attracts no agency fees, using social pathways can be a cost-effective and targeted addition to a recruitment toolkit, but it can also be intensive if researchers don't know what to do. So that researchers don't need to constantly reinvent the wheel, you should provide them with the following guidance:

- The approval to recruit socially so that researchers know that they've got a green light (or not). People don't want to do wrong, so help them by being explicit.
- Protocols that state whether and how to gain approval from partners like marketing or legal.
- Tools and protocols for requesting and collecting *personally identifiable information (PII)* when recruiting via social channels.
- Branded collateral or a style guide to use in creating social media assets.
- You might build ongoing partnerships with invite-only forums or fan clubs for researchers to leverage.
- A list of social channels to consider, the cohorts they might find, and the pros and cons of each.
- An internal forum, say a Slack channel, to share good and bad social recruiting experiences.

You should also consider the journey of participants who are recruited via social channels. If you manage an in-house recruitment panel, you could invite them to join via a friendly message attached to their participant thank-you gift.

> **NOTE** **RECRUITING RESPECTFULLY**
>
> Research involves collecting a significant amount of personally identifiable information (PII), and recruitment is no exception. To enable targeting and screening or to build an in-house participant panel, you'll need to collect and retain PII. It's important that people's data is treated in ways that are both ethical and compliant with data privacy policies across the recruitment workflow. Getting this right is a significant part of delivering research operations. You'll learn more about research ethics and data compliance in Chapter 9, "Respect in Research."

Recruiting Colleagues

Several years ago, I was asked to chat with a product manager about my experience using a company tool. The next day, I discovered that the call had been recorded and was being presented as research internally! Because colleagues may inadvertently risk their reputation or their job during an internal interview, extra care must be taken to protect participants' data and their identity—and not cause stress. If there's a good reason to enable internal recruitment (it has plenty of caveats, which are listed below), you should do the following:

- Partner with your people team to make sure they're onboard with internal recruitment protocols.

- Work with human resources and the legal team to devise an employee-specific informed consent.

- Provide rock-solid data governance protocols to make sure that this sensitive PII is properly handled.

- Give internal participants an appropriate route to ask for their data to be deleted permanently (right to be forgotten [RTBF]).

- Provide training on how to recruit internal participants and, crucially, *when* to recruit internally. It may seem easy, but it's not always best.

- Provide an appropriate way to say thank you to employees for sharing their time and information. Egift cards aren't always appropriate for employees, but swag or a charity donation might be.

- Finally, if there's a genuine and regular need for internal participants, you might build a panel of employees who are keen to take part in research.

When you consider the effort involved in doing internal research ethically, it's rarely necessary to recruit colleagues for anything other than real-deal internal research. This is especially true in a world where there are plenty of alternative options for recruiting public participants. Also, within a democratized research environment, the perceived cheapness and speed of recruiting colleagues can create an environment in which people rarely reach out to actual customers, which can weaken the veracity of research.

It's common for ResearchOps to do internal research with researchers to improve their experience. But particularly as the research team grows and becomes less tight-knit, researchers might simultaneously feel worried that they'll be compromised and yet awkward about turning a request down. It might feel strange taking informed consent from your colleagues, but by doing things formally, you'll alleviate any worries that researchers might have and mitigate any risks to the organization. Plus, it's good to practice what you preach! If you want to understand more significant operational themes, you're better off hiring a research agency to do the research for you. Outside of data privacy, this approach also gives researchers the opportunity to speak freely.

Each of these participant resources demands a significant amount of energy to mobilize and maintain, so it's vital to make sure that the resources you choose are in direct support of your participant recruitment strategy. For each common cohort that you define, resource that you provide, and discipline that you provide it to, you'll need to design an often unique or nuanced set of logistics that researchers should use to engage—not just as researchers or company ambassadors, but as fellow human beings.

Address the Ten Steps of Participant Recruitment

The style in which researchers find and communicate with potential participants is intrinsic to the participant's experience not just of the research, or the researcher, but also the brand. Depending on the needs of the cohort and the platform, the style of communications can change quite significantly. It's important to help researchers navigate these courtesies. To achieve this scalably, you'll need to provide flexible standards—recall Deming's bricks, lumber, and sandwich loaves—that researchers can use repeatedly to achieve these key steps (see Figure 5.5):

1. Write a recruitment brief.

2. Source participants.

3. Target and screen participants.

4. Invite participants to take part.

5. Schedule participants.

6. Invite research observers.

7. Practice informed consent.

8. Say thank you.

9. Data cleanse when the study ends.

10. Offer ongoing support.

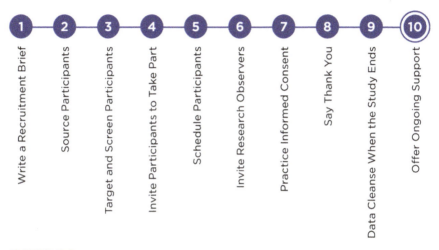

FIGURE 5.5

It's often assumed that recruitment ends when a participant is scheduled, but there are 10 key steps to participant recruitment, the final being continuous.

Write a Recruitment Brief

Of all the steps of participant recruitment, this is the lightest to support operationally. Paradoxically, it's one of the most important from a craft perspective because it sets the strategy for a successful recruitment. Researchers should always write a formal recruitment brief to plan their approach, and describe their participant needs to an external agency or an in-house recruitment team. Here, operations can support researchers in brainstorming their approach, and provide them with standardized request forms to make sure that projects are set up for success. (See Chapter 7, "Seamless Support.")

Source Participants

The topic of sourcing participants has already been covered in depth, but it's worth restating it because it's typically *so* tricky for researchers to find the right people. It's a huge operational win if you can make this step easy.

Target and Screen Participants

Targeting describes accessing and filtering a population so that it meets the recruitment criteria. *Screening* describes surveying the target population to refine the list of potential participants and understand who's interested in taking part.

A cohort-orientated approach will provide a way for researchers to broadly target the right audience, simply by way of the recruitment tactic they choose. But they'll often need to refine their search by targeting within that cohort. Agencies will target and screen for researchers, and all recruitment platforms feature targeting and screening capabilities as standard. But if researchers are recruiting via social channels (a form of targeting in itself—again, think TikTok users versus LinkedIn), they'll need a stand-alone tool to screen potential participants.

Screening can be done using a survey tool, but a survey tool can cause all sorts of problems if put in the wrong hands—or too many hands. Outside of the plague of poor survey design (cue Caroline Jarrett's Rosenfeld Media book, *Surveys That Work*), if an audience is oversurveyed, overemailed, or overintercepted, response rates can fall. "Overfishing" in one part of the business, say product marketing, can impact response rates in other parts of the business, too.

If you do offer a survey tool to support screening, put governance in place to manage access, permissions, frequency, and survey quality, just as you would for intercepts. A better solution is to provide researchers, and especially people who do research, with screening capabilities that are built into recruitment-specific tools where they are less likely to be abused. It's also common to create standardized screener templates that researchers can adjust to suit their scenario. Make sure to align these with your common cohorts.

TIPS AND TRICKS FOR SUCCESSFUL SCREENING

Writing successful screeners is a skill in itself. Experienced ops professionals and researchers build various tactics into their workflows, like prescreeners, participant prework,[11] and red herrings[12] to help them spot participants who fit—and don't fit—the bill. To brush up on these skills, it's useful to read books about survey design. You should also read Chapter 2, "Research Logistics" in Steve Portigal's book, *Interviewing Users,* 2nd ed. Steve covers participant recruitment from the craft point of view.

Invite Participants to Take Part

To help researchers easily invite, schedule, and give participants important session information, you must provide researchers with branded and customizable communications templates that are ideally built into the recruitment workflow. The most common communications templates are:

- **An invitation to take part in a session:** This should include details about the session and instructions on how to schedule a slot. A researcher will send this to screened participants who have put their hand up to take part.

- **A participant information sheet:** Once the participant has booked a slot, send them a booking confirmation that includes clear and concise instructions about what to prepare and expect on the day, along with a copy of the consent form.

- **An invitation to observers:** A message that can be sent to stakeholders to invite them to observe a session. It should include session details and observation dos and don'ts. You'll learn more about observer etiquette under the heading, "Invite Research Observers."

---

11 In his book, *Interviewing Users,* 2nd ed., page 56, researcher Steve Portigal says about participant prework: "Assigning prework can reveal recruited participants who are too busy or not fully engaged. If they don't respond, or respond late, or respond without much detail or insight, you can prioritize other participants for follow-up interviews." It's a great tip.

12 A red herring in a screener survey is a false bit of information, say a made-up product or fact, that recruiters can use to spot ill-informed (or dishonest) participants.

- **A participant reminder:** A brief note about the session and how to attend. This should be sent to participants a day or two before the session and can help reduce no shows. It can also help researchers plan a replacement if the participant cancels on them.

A PARTICIPANT INFORMATION SHEET

The most important asset a researcher needs as part of the recruitment process is a customizable information sheet. The participant information sheet should be sent to the participant when the session is scheduled, in order to:

- Confirm the session time and location. Include the time zone, if appropriate.
- Let the participant know if they should prepare anything for the session. For instance, should they have their mobile phone available to use?
- Provide a preview of the informed consent agreement, which they'll be asked to sign at the start of the session. This allows participants time to read the consent in detail, ask questions ahead of time, or cancel the session should they choose.
- If the location is a physical space, like a research lab, include advice about finding the location and transport and parking options.
- Is the location accessible? This is important in remote video contexts, too.
- Let the participant know whom to contact should they get lost or delayed.
- If the session is being held at the participant's home, let them know who to expect to arrive. It can be off-putting if a gang arrive when you expected just one person.
- State the agreement regarding their thank-you gift and when and how they should expect to receive it. (This should be included in the consent form, too, but it's worth repeating.)

Depending on the recruitment workflow and your technology, you might send the participant a customized information sheet automatically as soon as they book a session. If you do this, make sure to send the researcher a copy so that they know what's been communicated.

Consistently branded communications throughout the recruitment workflow help participants to build trust before their research session, which is invaluable. Assets that are even slightly off brand can throw up red flags for potential participants, and even confirmed participants, who might think your invitation is phishing or spam. Work with your organization's brand or marketing experts to make sure that your assets are spot on. Also, unless research topics are always quirky and fun, aim for a neutral tone. If researchers are exploring a sensitive topic, they won't want to use a template that's inappropriate: "Hello! We'd love to talk to you about your recent car accident." It's dark humor, but you see the point. Finally, the tone and style of communications can vary wildly depending on the audience, so think carefully about the audience's expectations and relationship with the company.

Schedule Participants

Scheduling can be one of the most painful and admin-heavy parts of the recruiting process. Recruitment platforms always include scheduling capabilities, and agencies will handle scheduling for you, but if a researcher needs to recruit off the beaten path, they'll need this important capability. Tools like Calendly and YouCanBookMe are perfect for this situation, but, because they tend to require access to an organization's calendaring system to work, they can be tricky to get through procurement and security. Still, if you don't already have one available, they're worth the effort. If you do provide a scheduling tool, make sure to provide templates and protocols for how to use calendar apps so that they feel on-brand and integral from the participant's point of view.

> **NOTE** WHY NOT JUST USE AN EMAIL AND CALENDAR APP?
>
> As a rule, researchers shouldn't use email and their standard calendaring app to communicate with participants and schedule sessions. Emailing back and forth to agree to a date is unnecessarily admin heavy, but, even more importantly, it also means that PII will be strewn across calendars, email, and local drives and therefore hard to find should someone request that you delete their PII.

Invite Research Observers

There's nothing more powerful than seeing something for your-self. In support of research knowledge management, which will be covered in Chapter 6, "Long Live Research Knowledge," you'll need to enable researchers to easily invite people to observe sessions while keeping PII private. It's important that researchers don't invite observers using the same calendar invite that was used for the participant, unless invitees' details can be hidden. A hacky work-around is to instruct researchers to create a duplicate calendar invite for their observers only, which isn't particularly efficient. Instead, you can automate the duplication of the calendar invite using a tool like Zapier or manage the workflow via a recruitment or research moderation platform.

STANDARDIZE OBSERVER ETIQUETTE

It's a good idea to provide both researchers and observers with training about observer etiquette. The training should include information about how to:

- Respect participants' privacy before, during, and after the session.
- Communicate with the research moderator during the session.
- Take part in ways that are useful, such as a standardized format for how verbatim quotes should be written on sticky notes or in a note-taking tool.

But don't make this a one-off; it's not bad practice to remind someone of the protocols before every session that they observe. I've known researchers to ask observers to sign an etiquette agreement before every session. If this approach feels right to do, and it's working for researchers, you might make this approach standard.

Practice Informed Consent

Informed consent should happen in *every* research session, never mind how someone was recruited or who they are. Critically, informed consent involves more than just getting a form signed. Instead, as the name suggests, the participant should feel informed about their legal rights, how they'll be treated, and how their data will be used both during and after the session. (See Chapter 9.)

Practicing informed consent should be a thoroughly designed element of your participant recruitment workflows. It should be simple enough that it becomes habitual to researchers, thorough and understandable enough that it achieves the goal of informing participants, and short enough that it doesn't "steal" too much time at the start of a session. In all circumstances, you'll need to provide researchers with:

- Training on how to practice informed consent, what it means, and why it's important.

- A modus operandi for sharing the consent form with their participants prior to the session. As mentioned earlier, this could be delivered as part of the session information sheet. (Depending on your legal requirements, you may still require that participants sign the form during the session.)

- A bullet-point script to help researchers remember salient points when taking consent.

- A route for participants and researchers to get support post-session should a participant want to amend their consent or exercise their right to be forgotten (RTBF).

There are many tools that can help you manage informed consent efficiently and compliantly, and leaning into your organization's existing tools and protocols will save you both time and money. But if you want something more custom, consentkit.com has been built with ResearchOps in mind, and it now offers "responsible recruitment" services, too.

Say Thank You

The practice of rewarding people for taking part in research has a long-standing history that spans at least 100 years.[13] In the user research world, saying thank you to participants for their time and information usually includes giving participants some kind of gift. Many researchers call thank-you gifts "incentivizing," because they're incentivizing participants to sign up to take part in the study. Some researchers even say that they're "paying" the participant. While this is true, for legal reasons that will be explained in a moment, it's best to avoid using words like "incentivizing" and "paying." A thank-you gift could be:

- Money in the form of cash, an egift card, or prepaid balance (digital or physical)
- A charity donation
- Branded swag, such as a journal, baseball cap, or t-shirt
- Access to a product or service for free
- A prize that's won via a research sweepstake
- The satisfaction of having impacted a roadmap

> **NOTE** **LEGAL RESEARCH SWEEPSTAKES**
>
> A research sweepstake offers participants the chance to win something by taking part in a study—usually a survey of some kind. In the United States, sweepstakes are regulated by the Federal Trade Commission, which oversees advertising and unfair trade practices. In most states, sweepstakes are allowed as long as participants aren't required to pay to submit an entry. Different jurisdictions have different laws about sweepstakes, and it's important that they are clearly set out as part of an informed consent. Before offering researchers sweepstakes as an incentive option, check in with your legal team.

There are legal implications around thank-you gifts, which are important to understand, particularly if you're thanking participants in large numbers. Here's what you should watch out for:

- Stating that you will "pay someone" implies that an organization is employing someone to take part in research, which misconstrues the relationship both philosophically and contractually.

---

13 Christine Grady, "Payment of Clinical Research Subjects," *The Journal of Clinical Investigation* 115, no. 7 (2005): 1681–1687, https://doi.org/10.1172/JCI25694

- In the U.S., research thank-you gifts are taxable income. According to the Internal Revenue Service in the U.S., if an organization gives someone more than $600 in a fiscal year, they're required to send the recipient a Form 1099-MISC.[14] When you set up your thank-you gift service, work with your legal team to ascertain what rules apply in your context.

- As noted previously, advertising and trade regulations apply to sweepstakes or competitions, so make sure that your practices are legal before offering them.

These days, there are excellent tools in compliance with regulations, which support thanking participants at scale such as egift card platforms and swag box outlets. I've used Tremendous for egift cards and SwagUp for swag in the past, but there are others like Tango and BHN Rewards (formerly Rybbon) that are equally popular. When you're working across multiple countries, make sure to consider the impact of currency rates, standard wages, and how that translation impacts your recruitment outcomes. In developing countries, a $100 thank-you gift would come with a lot more weight than in Luxembourg, for instance. You'll also need to consider standards around types of participants, their contexts, and the organization's relationship with them: high-paid executives may be insulted by a $100 egift card, whereas a standard consumer is likely to be thrilled for the extra cash.

Data Cleanse When the Study Ends

Across the participant recruitment workflow, researchers will collect PII. When people sign up to a panel, complete a screener, schedule a session, and accept a thank-you gift, they will likely share personal data. In *all* these instances, people should consent in one way or the other to sharing their data. In some cases, consent might be specific

---

14 "About Form 1099-MISC, Miscellaneous Information," *IRS*, www.irs.gov/forms-pubs/about-form-1099-misc

to the context, while at other times it's captured under a previously agreed consent or under the organization's broadly applicable privacy policy.

It's important that your operations group puts protocols in place not just to safeguard data captured during participation, but also to safeguard and govern data captured during participant recruitment. As part of this, researchers should be encouraged to *data cleanse when the study ends*—a nice little mantra. They should clear temporary data from email clients and local drives (ideally, it's not stored locally in the first place, but however hard you work, no process is watertight), and move data that's legal and useful to keep to a properly governed data storage tool.

Offer Ongoing Support

There are all sorts of questions and concerns that past or potential participants might have: "Where is my thank you gift?" "The egift card has expired." "I would like to exercise my right to be forgotten." "I haven't received an invite to take part in research, have I done something wrong?" And even, "I was unsettled by the research experience." Participants' research experience is key, so you'll need to provide them with a way to get in touch. You should also provide researchers with standardized (or better yet, built-in) ways to let people know that this support is available. You might include a note about follow-up support in thank-you gift messaging; you might also request feedback about their research experience. Ideally, your participant support will integrate with your organization's existing support channels so that participants don't need to seek you out via unique communications channels. (See Chapter 7.)

> **NOTE THANKING DISHONEST PEOPLE**
>
> However sharp your recruitment is, on occasion, a dishonest participant might still succeed at attending a session. Even though participation policies tend to state that thank-you gifts will only be awarded to genuine candidates, it's common practice to send the thank-you gift even if the session was not useful. Doing anything else simply isn't worth the hassle or potential brand damage.

Make Recruitment Inclusive by Design

In her book, *Invisible Women: Data Bias in a World Designed for Men*, feminist Caroline Criado Perez investigates gender inequality in how the world is researched and then designed predominantly for men. From medicine to mobile phones, and car safety to public transport, women are systematically excluded from research. Operations has a weighty obligation to set inequality right in its own quarters. You can do this by training researchers to recruit inclusively and providing them with the participants and resources they need to make good intentions real, which is particularly relevant if you manage an in-house panel. Your participant recruitment practices should foster diversity across age, disability, gender reassignment, marriage and civil partnership, pregnancy and maternity, race, religion or belief, sex and sexual orientation. Additionally, your research practices and infrastructure should support income diversity, too.

Your context might impact your ability to meet ultimate goals of equality. If research focuses on motor sports, for instance, inequalities within the sport will impact your recruitment diversity. Even so, there's always a reason to invest in inclusivity. As Maya Angelou said, "Do the best you can until you know better. Then when you know better, do better." There are a number of tactical moves you can make to improve inclusivity in research participation:

- Make sure that your panel populations are diverse through demographic analysis and targeted marketing.
- Challenge the external agencies you choose to work with on their equality practices.
- Ensure that your participant experience is accessible: people living with a disability should be able to take part equally.
- Make sure that content is written in plain language so that people from all walks of life can contribute.
- Where appropriate, allow participants to take part in research using either iOS or Android-operated devices, the latter being generally more affordable.

In a Nutshell

Participant recruitment is *so* big and detailed that keeping view of the forest for the trees can be tricky. It's easy to get stuck in the detail of running a panel, procuring a platform, or building templates, only to lose sight of the wider system and its purpose. For this reason alone, it's crucial to devise a recruitment strategy that outlines an approach for providing researchers with:

- **Standardized populations, templates, and tools, which they can adapt to recruit the participants they need.** Common cohorts should offer a doorway into more discrete populations, but cohorts should not be so broad that they lose all definition. "Customers," for instance, isn't a cohort no matter how excellent your filtering systems are.

- **People-centric ways to know where to find common cohorts, like a map or a decision tree.** On the same note, "XYZ agency" doesn't let researchers know who they will find, whereas the addition of "XYZ decision-makers" or the context of a decision-tree does.

- **Specialized recruitment operations that align with organizational priorities: car-share customers if you're Uber, perhaps.** You should also provide specialized ops to support ethical needs, like recruiting people from marginalized groups.

- **Distinct methods for accessing populations in ways that are appropriate to the cohort, the recruitment tactic, and the needs of the research.** For instance, a discrete panel that uses highly personalized communications to engage with VIP customers, or intercepts to engage people at a particular point in a transaction.

- **Protocols and assets suited to each recruitment platform and cohort.** These can help make the 10 steps of recruitment as seamless and respectful as possible for participants, researchers, and internal partners like marketing or sales.

CHAPTER 6

Long Live Research Knowledge

More than thirty years ago, Peter Drucker wrote an article for the *Harvard Business Review* called "The Coming of the New Organization."[1] In it, he predicted that at the turn of the century "The typical business will be *knowledge-based*, an organization composed largely of specialists who direct and discipline their own performance through organized feedback from colleagues, customers, and headquarters." Drucker made specific mention of the research profession, too, predicting that with the information age, the sequential order of research, development, manufacturing, and marketing would be replaced with what might today be colloquially called "team sport": specialists from multiple fields working synchronously to deliver a product or service from inception to marketplace.

Drucker's words couldn't have been more true—and here we are in the thick of it. This fundamental shift has seen the rigor and routine of the assembly line give way to a new way of working, one in which knowledge is broadly distributed and everyone must "think through what information is for them, what data they need: first, to know what they are doing; second, to be able to decide what they should be doing; and finally, to appraise how well they are doing." It's within this context that the user research profession has grown up. It's within this context, too, that as "knowledge excavators" and "merchants," research has the potential to become key to the success of the largest organizations of our time—*every* organization of our time. Despite economic ups and downs, the stage is set for research to shine. But only if researchers, in tandem with ops, become masters not just of creating the right knowledge, but also organizing and distributing it, and helping people to derive *insight* from it, too.

FOOD FOR THOUGHT
INTERROGATING "INSIGHT"

Insight—it's a word that's become nearly cliché in its overuse, both within the user research profession and the broader user experience field. Yet prod for a shared understanding of what "insight," or more specifically "research insight" means, and you'll find that the response is variable, even within research teams.

---

1   Peter F. Drucker, "The Coming of the New Organization," *Harvard Business Review*, January 1988, www.hbr.org/1988/01/the-coming-of-the-new-organization

In a blog post, The New Zealand–based agency, We Create Futures, shared this framing:[2] "An insight is a compelling truth, narrative, or story, created in the analysis and synthesis of multiple data-points. Its role is to inspire or reveal a new perspective, opportunity or mind-shift in the reader or viewer." This definition is *so* good because it speaks to the importance of an insight being more than just a "nugget"[3] or a report (a noun); it should shift or enrich how someone thinks.

The Promise of Well-Managed Knowledge

Most researchers don't need convincing that they'd reap benefits from managing research knowledge better; outside of fixing participant recruitment, it's often a top priority for research operations. What surprises most people is just how much good knowledge management has to offer. Beyond answering that most common of questions, "What do we know about XYZ?," well-managed knowledge can help a research organization do the following:

- Showcase the accumulated weight of the research team's contribution to organizational knowledge and help map its impact, which should be guided by strategy. And if you're not making an impact, it can help you understand why.

- Deliver speed and efficiency to the organization by using previous research to provide faster and more cost-efficient answers or hypotheses to new research questions. However, there's a kickback: This benefit can also quell the organization's desire for original research, so watch this balance.

- Drive strategic change by providing the capability for richer original research and secondary research. A library will deliver its weight in gold if you make secondary research a strength.

- Foster strategic partnerships and break down silos by supporting knowledge sharing and learning within and across teams. Sharing is a great way to make friends.

---

2   Chris Jackson, "The Importance of Customer Insight," *We Create Futures* (blog), www.wecreatefutures.com/blog/the-importance-of-customer-insight

3   "What Are Atomic UX Research Nuggets?," User Interviews, www.userinterviews.com/ux-research-field-guide-chapter/atomic -research-nuggets

- Support research planning and strategy by revealing current and potential future knowledge gaps or unnecessary research repetition: reducing duplication of research is a key knowledge management theme.
- Retain knowledge within the organization when people change their focus or role or leave permanently.

BUILD AN OPS TEAM

PROMISES MADE REAL

While working at Atlassian, I hired an ingenious[4] librarian to deliver a research library, and it's paying off. Alison Jones (that librarian) shared this story in a blog post: "A lead researcher received a request from a stakeholder regarding what we knew about an issue where the answer lay in content spread across a number of reports."[5] A library meant that they could "gather up the relevant content, give that content to her team and ask them to synthesize from the content provided. Before the library existed, gathering such content might have taken a couple of weeks." This sort of result takes specialist skills and investment, but the return on investment is worth it.

The benefits of well-managed knowledge extends beyond outright strategic value. By making the most of what you know, researchers (and their managers) may be better supported in their work, too. Well-managed knowledge can:

- Help new team members and collaborators hit the ground running by giving them an overview of what's already known about their topic of interest, and *who's* in the know. This can help newcomers build a useful network from the get-go.
- Give researchers the opportunity to store insights not related to their research study but which might have value in the future.

---

4   When I set out to hire a Senior Research Librarian for Atlassian, a company I left in early 2023, I learned that finding a librarian who has built libraries from scratch and could "eat taxonomies for breakfast," per the job ad, was more challenging than you might think. Many librarians end up in clerical roles within libraries, so keep this in mind when you're hiring.

5   Alison Jones, "Unlocking the Power of Research: The Atlassian Research Library," *Medium (blog)*, March 1, 2024, https://medium.com/@ajones6_48278/unlocking-the-power-of-research-the-atlassian-research-library-eec30d83614b

This is especially true when it comes to communities of practice or "knowledge campfires." (See "Defining an RKM Strategy.")

- Help researchers grow their reputation as a pundit on certain topics, bringing a greater sense of meaning and richness to their work, assuming their environment supports that kind of focus. In reality, researchers are often moved from one topic or team to another, leading to a shallowness in what they know. A library can be a valuable mitigator (and illustrator) of this kind of knowledge and capability debt.

- Give a researcher and their manager the opportunity to review their body of work when preparing for performance conversations or promotions. A note of caution though: research impact can be *tacit* and *cumulative*; it can take months or years for research to show overt impact. Besides, producing content isn't the only sign of productivity or success in research.

Well-managed knowledge can also support ops with the capability to do the following:

- Track and delete data associated with a participant should they request the right to be forgotten (RTBF). RTBF is a requirement of the EU's General Data Protection Regulation (GDPR), a regulation that's seen as the gold standard of privacy in the digital age and is being emulated by countries worldwide. (See Chapter 9, "Respect in Research.")

- Monitor and communicate research engagement by providing metrics around things like content views, downloads, comments, and more. (See Chapter 10, "Money and Metrics.")

Mention the term *research knowledge management,* and your instant association was likely a repository or library. Libraries and repositories are core knowledge management assets, for sure, but there are several more tactics to consider. This chapter covers the key concepts that underpin knowledge management (KM) in the wider world, and what you should consider when you devise a research knowledge management (RKM)[6] strategy, specifically. But first, there's something that cannot be ignored: countless well-intentioned, not to mention expensive, repositories and libraries bite the dust, and it's useful to understand why.

---

6 RKM is an acronym that I've made up, and which I'm introducing here.

UNDERSTANDING KEY KM TERMS

Even within the information science, librarian, and knowledge management professions, the jury is still out on the meaning of key words and terms, including substantive words like *knowledge*—a philosophical wormhole. But there are broadly held views that are good to honor. *Library*, *repository*, and *archive* are a good place to start because they're often confused in user research, even by popular research repository tools. *Taxonomy*, *tags*, and *folksonomy* are also a pertinent trio because they're often confused and misused, too, within the field of user research.

A **library** is an organized collection of published materials that can be accessed by others en masse. Crucially, a library must function as a consumer-facing service as much as a collection-management instrument. A research library should catalog and make available research reports or other consumer-ready assets, like presentation decks or research share-back videos. A library must be centrally controlled.

A **repository's** focus is on safeguarding and preserving assets. Assets should be findable—the repository would be useless otherwise—which means that it must house a robust *taxonomy*. But assets needn't be published or made publicly available as in the case of a library. In the world of research, a repository might catalog raw insights, consent forms, raw video or audio files, photographs, transcriptions, or items picked up during field research. You'll likely use a repository to store materials as part of data governance.

An **archive** is generally associated with storing and safeguarding historical documents. The word *archive* is generally considered to be interchangeable with *repository*.

Taxonomy[7] is a derivative of the Greek taxis (arrangement) and nomos (law).[8] In simple terms, it's about making information easy to find, even within a large and diverse collection. It's the science of

---

7   Carolus Linnaeus is regarded as the founder of modern taxonomy, though, as a naturalist, he developed the concept of taxonomy to place living organisms "into a series of categories according to their presumed natural relationships, with kingdoms divided into phyla, phyla into classes, classes into orders, etc." Quote from the Merriam-Webster blog post, "'Folksonomy': Let's Get Organized," see footnote 9 on page 131.

8   A. J. Cain, "Taxonomy," *Encyclopedia Britannica*, December 9, 2022, www.britannica.com/science/taxonomy

classification: the arrangement of superior and subordinate groups of things that are governed by a centrally controlled and hierarchical structure and vocabulary, usually defined as *metadata*. Confusion abounds about what a taxonomy is—and what it is *not*—and it's worth clearing that up. A collection of tags or categories isn't necessarily a taxonomy. For instance, the fields that you might use to define the information architecture of a library such as "researcher," "product," "research topic," and so on are not a taxonomy. Instead, the taxonomy is the controlled, and often complex and constantly growing, language that populates those fields, like "Jane Smith" under "researcher," "Lingo Pro" under "product," and "Freemium" under "research topic." Taxonomies aren't assets that can be adopted or lifted and shifted from place to place. To succeed, a taxonomy must be custom fit and constantly controlled, which is a highly specialized skill best handled by a knowledge professional.

Folksonomy is a term that was coined by information architect Thomas Vander Wal in 2004[9] to describe the spontaneous and social classification methods people used to make content findable across social networks—like hashtags on LinkedIn, Instagram, or Twitter. Critically, a folksonomy is an evolving and unregulated type of metadata. Per Merriam Webster's informative article: "It only needs to make sense to you." This is in contrast to a taxonomy which is centrally controlled.

Tags are discrete keywords or terms. It's a kind of metadata that describes an item in shorthand allowing it to be found again, like "strawberry" or "banana," if you were to tag a supermarket's grocery aisle and, in particular, the "fruit" aisle. Tags can form a controlled vocabulary, as in the case of a taxonomy, or if uncontrolled and used socially, they can form a folksonomy. A collection of tags isn't necessarily a taxonomy, which is a common confusion.

A successful KM system will usually make the most of folksonomies to support communities of practice and taxonomies to support reliable knowledge services, for each is of benefit if used in the right context. (You'll learn more about communities of practice and knowledge services under the heading "Defining an RKM Strategy.")

---

9   "'Folksonomy': Let's Get Organized," *Merriam-Webster* (blog),
    www.merriam-webster.com/words-at-play/folksonomy-tagging-social-media

Why So Many Efforts Fall Apart

If you watch sci-fi—I'm thinking about the *Foundation*[10] series, but *Star Wars* works too—you'll know this image well: a space craft from eons ago lies derelict at the bottom of the ocean waiting to be brought back to life, and countless more are scattered across the universe. Unfortunately, it's an image that chimes well with the state of knowledge management in the user research profession. The profession is a graveyard of libraries and repositories with innumerable insights lost within them. Even though industry-specific tools are popping up like mushrooms and being enthusiastically procured, and people have hacked various productivity tools like Jira and Airtable, there are few *sustained* success stories to lean into. There are several reasons for this, and they are the following:

Efforts often aren't strategic. There's a tendency to rely on just one tactic such as a library to do it all, but it won't work. It's also not uncommon to procure a tool without considering the metrics for success or the *specific* purpose; I call this *knee-jerk operations*.

Accumulation isn't designed for. Knowledge assets, such as research reports, data, or insights, tend to accumulate at a rapid pace—quicker than you might think. If a team of 40 researchers produces 40 reports a month (and doesn't scale), you'll be managing access and archiving for 1,140 research reports within three years, plus thousands of related data assets. Yet few efforts include planning around how to manage a constantly growing, and aging, research collection.

There's limited understanding or respect of long-established fields. There's a wealth of knowledge and skills to lean into (and hire, if you know whom to look for), and no need to reinvent the wheel. Everything that research practices want from knowledge management has already been delivered in countless contexts for hundreds of years, from libraries in ancient Greece to digital libraries and state libraries today and scaled-up knowledge management efforts in academia and enterprises. The opportunity is *not* to invent new ways of doing things, rather, it's to adopt and adapt the wealth of expertise that already exists.

---

10 *Foundation*, Apple TV+, https://tv.apple.com/us/show/foundation/umc.cmc.5983fipzqbicvrve6jdfep4x3

User research-specific tools aren't always fit for purpose. The siloed view of RKM means that research-specific tools aren't being built using tried-and-tested information sciences frameworks, which is leading to a fundamental lack in capability. In several cases, "repositories" aren't manageable at even moderate scales because the capability needed to manage a structured taxonomy, the backbone of any scalable repository or library, is nonexistent. These tools *are* useful, but as communal "campfires" not libraries (see "Communities of Practice"). When a robust library tool is required, there are countless available from outside of the user research profession, and they are often vastly less expensive.

Organizations aren't hiring the right people. Knowledge management needs long-term, constant, and *skilled* attention, yet few organizations have invested in hiring a full-time research librarian or similar—even a full-time data gardener is a step in the right direction. Just as research involves much more than just chatting, librarianship involves more than just tagging things. So, while it's possible to deliver a degree of knowledge management without a librarian, there's no getting around this universal truth: You get what you put in.

Taxonomies are not being managed. To scale research knowledge management, you will need a structured taxonomy, which means that you'll need to work with someone who knows what they're doing. (I'm banging the same drum, but it's worth repeating.) There's also often a lack of planning around how to mitigate and manage the taxonomic mess of a folksonomy that has had its limits stretched. Folksonomies don't scale well beyond a small group of well-coordinated and orderly researchers or a small collection of knowledge assets. Even then, time, scale, and change tend to erode collective intelligence and consistency (and folksonomies and taxonomies).

If you take these fundamental pitfalls into account, your chances of delivering a successful knowledge management system will upshift, no doubt. But to build a souped-up system of any kind, it helps to get under the hood to understand how everything works. In the world of knowledge management, this means understanding the basics of how people think, and how they learn.

A FUNDAMENTAL CHECKLIST OF TO-DOS

It's worth transforming these common pitfalls into a checklist of fundamental to-dos:

- Do you have a *knowledge management strategy* in place? You'll learn how to approach a knowledge management strategy throughout this chapter but specifically under the heading, "Defining an RKM Strategy."

- Have you forecast, as best you can, how your research collection will scale over several years and what will comprise your collection? This will give you a grounded sense of how much content you'll need to input, maintain, and archive over time, which will define both design and resourcing.

- Before procuring a tool, have you considered whether it supports your strategy, whether it will handle growing numbers of people accessing it, as well as a constantly growing research collection over several years?

- Have you learned about or accessed skills from the fields of librarianship, knowledge management, or information science? These fields specialize in the collection, classification, analysis, manipulation, storage, distribution, movement, and protection of information, which is skillful work.

- Do you know how you'll manage the various types of metadata needed to keep information organized, accurate, and findable? Findability predicates trust, which is a weighty theme.

- Have you hired a specialist to design and look after the systems and services in the long term? As mentioned, I've hired a Senior Research Librarian in the past and would do so again.

- Even the best-built library tended to by the most skilled knowledge manager will come to nil if the library and its services aren't well adopted. Do you have a plan in place to handle communications, support, and change management?

Must-Know Theories, Models, and Memes

One of the most popular notions about knowledge management is the DIKW hierarchy, or the data-information-knowledge-wisdom hierarchy (see Figure 6.1). You'll likely recognize it in an instant because it's meme-like in its popularity for explaining the data-to-wisdom journey. The hierarchy was made popular by Russell Ackoff, a pioneer in the field of operations research, systems thinking, and management science, in his address accepting the presidency of the International Society for General Systems Research in 1989. But it was Milan Zeleny, a Czech American economist, who first detailed the hierarchy in his 1987 article "Management Support Systems" in which he equated data, information, knowledge, and wisdom with various forms of knowledge: "know nothing," "know what," "know how," and "know why," respectively.[11]

FIGURE 6.1
The data-information-knowledge-wisdom hierarchy (DIKW) is often used to explain the data-to-wisdom journey.

The DIKW hierarchy has become dominant largely because it makes the work of knowledge management seem so linear—even easy. But as philosopher-technologist David Weinberger wrote, "Knowledge is not a result merely of filtering or algorithms. It results from a far more complex process that is social, goal-driven, contextual, and

---

11 Nikhil Sharma, "The Origin of Data Information Knowledge Wisdom (DIKW) Hierarchy," *ResearchGate*, April 2008, www.researchgate.net/publication/292335202_The_Origin_of_Data_Information_Knowledge_Wisdom_DIKW_Hierarchy

culturally bound. We get to knowledge—especially "actionable" knowledge—by having desires and curiosity, through plotting and play, by being wrong more often than right, by talking with others and forming social bonds, by applying methods and then backing away from them, by calculation and serendipity, by rationality and intuition, by institutional processes and social roles."[12] You would be hard pressed to find a better quote to guide this important work.

If it's not possible to operationalize for a neat four-step DIKW process, the question is: Can you operationalize for Weinberger's nonlinear knowledge space? You *can*. But instead of seeking to control or wanting to measure knowledge, you'll need to deliver spaces and services that enable curiosity, plotting, and social interplay as much as controlled knowledge stores.

The first recorded appearance of the DIKW concept was in 1934 in a poem called "The Rock,"[13] written by T. S. Eliot:

"Where is the Life we have lost in living?

Where is the wisdom we have lost in knowledge?

Where is the knowledge we have lost in the information?"

And almost half a century later, musician Frank Zappa articulated the DIKW concept in his song, "Packard Goose."

A Philosophical Wormhole Condensed

When first approaching knowledge management, it can seem an excellent idea to define and understand what is meant by words like *knowledge* or *insights*. But even a rudimentary glimpse of these topics uncovers a philosophical world that is as equally fascinating as it is confounding and absorbing.[14] It is to enter a wormhole of philosophy and academia. You might ask: "Surely you need to know what 'knowing' or 'insight' means to deliver knowledge management?"

---

12 David Weinberger, "The Problem with the Data-Information-Knowledge-Wisdom Hierarchy," *Harvard Business Review*, February 2, 2010, www.hbr.org/2010/02/data-is-to-info-as-info-is-not

13 Nikhil Sharma, "The Origin of Data Information Knowledge Wisdom (DIKW) Hierarchy," *ResearchGate*, April 2008, www.researchgate.net/publication/292335202_The_Origin_of_Data_Information_Knowledge_Wisdom_DIKW_Hierarchy

14 Even Plato's proposal that knowledge is "justified true belief" was accepted by most Western philosophers until, in 1963, philosopher Edmund Gettier challenged the definition, and the debate has raged on ever since.

You don't. But a critical understanding of the dominant (or just plain relevant) theories of knowledge management and how people learn is essential. The three theories worth exploring are:

- The tacit dimension
- The SECI model
- Milton's six knowledge types

Each of these models has contributed to the development of a model for planning research knowledge management, which is called the *RKM Model*. You might to be tempted to skip the philosophical theories—if so, go ahead—but by taking the time to understand the genesis of the RKM Model, you'll be empowered to design and deliver a strategy that is custom-fit.

A WHISTLESTOP HISTORY OF KM

In 1993, CERN released the World Wide Web into the public domain and heralded the dawn of the information age. As a result, organizations everywhere were forced to reorganize to remain competitive in a world in which information was the currency du jour. It's within this context, and within the management consulting community at the time, that the now well-established discipline of KM was born. While the term *knowledge management* was, as the story goes, first used at McKinsey & Company in 1987 for an internal study on their information use and handling, it was Tom Davenport working at Ernst & Young, who, with this to-the-point one-liner, gave the discipline a face: "Knowledge Management is the process of capturing, distributing, and effectively using knowledge."[15]

Knowledge management brings together information science, library science, data management, information architecture, content strategy, social sciences, and more. It leans into any science or skill that will aid an organization in making the most of the know-how (*tacit knowledge*) and know-what (*explicit knowledge*) housed within their systems and people—just as ResearchOps works with a variety of professions and tunes them toward elevating research. You'll learn more about tacit and explicit knowledge under the heading, "The Tacit Dimension."

To expand your KM knowledge, follow Stan Garfield, Patrick Lambe, and Nick Milton—the latter's work is referenced in this chapter.

---

15 Michael Koenig, "What Is KM? Knowledge Management Explained," *KM World* (blog), May 4, 2012, www.kmworld.com/Articles/Editorial/What-Is-.../What-is -KM-Knowledge-Management-Explained-82405.aspx

The Tacit Dimension

Two decades prior to the emergence of the DIKW hierarchy, Michael Polanyi, a Hungarian-British polymath, devised the term *tacit and explicit knowledge*, which he explained in his book *The Tacit Dimension*.[16] *Tacit knowledge* is "learning by doing" or "know-how." As Polyani said: "We know more than we can tell." This type of knowledge is expressed through actions, like driving a car or riding a bicycle almost without thinking, and, because it's unconsciously internalized, it's difficult to articulate and document. Tacit knowledge represents people's experiences, intuition, and insight—i.e., beliefs, creativity, and feelings. It's the kind of knowledge that researchers aim to uncover in research interviews and hope that stakeholders will experience in engaging with their work. Tacit knowledge also means that research analysis and synthesis are hard skills to teach in any way other than shadowing.

Explicit knowledge, on the other hand, is "know that" (about the facts). Being explicit or unambiguous, it can be easily verbalized or codified: recorded using words, numbers, models, maps, decks, prototypes, audio recordings, research reports, and more. The fact that it can be codified means that it's also easy to communicate. It's also easier to capture, collect, and curate and, therefore, manage. Tacit and explicit knowledge are a binary concept, but, as professors Ikujiro Nonaka and Hirotaka Takeuchi sought to express via their *SECI model*, that's not the full picture. The SECI model provides an interesting and pragmatic application for RKM.

The SECI Model

In exploring the interplay between Polyani's tacit and explicit knowledge, professors Ikujiro Nonaka and Hirotaka Takeuchi developed the SECI model (see Figure 6.2).[17, 18] The SECI model illustrates how tacit knowledge is converted to explicit knowledge and back again via socialization-externalization-combination-internalization (SECI).

---

16 Michael Polanyi, *The Tactic Dimension* (Chicago: The University of Chicago Press, 1966).

17 "Nonaka and Takeuchi," Praxis, www.praxisframework.org/en/library/nonaka-and-takeuchi

18 Bandera, Cesar, Fazel Keshtkar, Michael R. Bartolacci, Shiromani Neerudu, and Katia Passerini, "Knowledge Management and the Entrepreneur: Insights from Ikujiro Nonaka's Dynamic Knowledge Creation Model (SECI)," *International Journal of Innovation Studies 1*, no. 3 (2017): 163-174, accessed June 6, 2024, https://doi.org/10.1016/j.ijis.2017.10.005

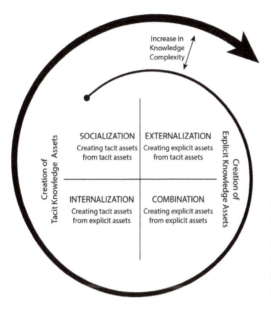

FIGURE 6.2
The SECI Model of
Dynamic Knowledge
Creation, adapted
from Nonaka and
Takeuchi's original.

- **Socialization** is the process of sharing knowledge. It describes the creation of new tacit knowledge by engaging with other forms of tacit knowledge, like listening, observation, conversation, or imitation and practice. Inviting stakeholders to observe research is a form of tacit knowledge sharing or socialization, as is sharing insights via a research presentation.

- **Externalization** converts tacit knowledge into explicit knowledge. This is what all researchers do in internalizing new knowledge gleaned from the process of doing research, and then externalizing it in the form of a report or deck, or some other form.

- **Combination** is the creation of further explicit knowledge by filtering, re-contextualizing, analyzing, and synthesizing already explicit knowledge sources like stored research data or reports. Secondary research or even analysis and synthesis, of course, are good research-world examples.

- **Internalization** involves converting explicit knowledge into tacit knowledge so that it may inform action. Internalization is what all researchers want their stakeholders to do so that they can make decisions based on the research, therefore delivering impact.

In 1998, CW Choo,[19] Professor in the Faculty of Information at the University of Toronto, added the concept of cultural knowledge to the SECI model. Cultural knowledge guides the norms that inform how a group of people respond to something new: it's the collective assumptions, values, and beliefs that "form the sense-making framework in which organizational members recognize the saliency of new information." For instance, if there's a general sense in the organization that researchers are disconnected intellectuals who live in an ivory tower, your research is less likely to be taken notice of, however excellent it is. In designing an RKM strategy, take time to consider the culture that the services will live within. As a gardener measures the pH of the soil before planting, so you should measure the pH of your organization before building.

On the other hand, being able to quickly find relevant content is key to building trust in content and, by proxy, trust in its producers. Colleen Jones, author of *The Content Advantage* and founder of Content Science wrote in a blog post: "We found that when people perceived content as hard to find, they also seriously doubted its accuracy, relevance, and usefulness to their goal."[20] So, building trust in research content requires a two-pronged approach: you must help people to easily access research that's relevant to them, and help researchers build relationships through genuine and consistent engagement protocols. (See Chapter 11, "Getting Priorities Straight.")

---

19 "Chun Wei Choo," University of Toronto Faculty of Information, www.ischool.utoronto.ca/profile/chun-wei-choo/

20 Colleen Jones, "Make Your Content Findable or Die," *Content Science Review* (blog), December 1, 2019, https://review.content-science.com/make-your-content-findable-or-die/

Milton's Six Knowledge Types

In his blog post "Tacit, Explicit and…What? Different Types of Knowledge, and the Definition Minefield,"[21] knowledge management author and consultant Nick Milton tackles the SECI model's vulnerabilities and offers this pragmatic advice: "Don't worry too much about definitions—recognize the different types (of knowledge) and how they must be addressed, and make sure your KM program addresses them all." His tactics can be usefully translated into knowledge types vis-à-vis tactics in RKM (see Table 6.1).

---

21 Nick Milton, "Tacit, Explicit and…What? Different Types of Knowledge, and the Definition Minefield," *Knoco Stories* (blog), October 15, 2020, www.nickmilton.com/2020/10/tacit-explicit-and-what-different-types.html

TABLE 6.1   KNOWLEDGE TYPES AND TACTICS IN RKM

Knowledge Type	Tactic
Knowledge that *can't* be expressed or articulated, variously called *tacit* or *implicit knowledge*.	It can be transferred through coaching and mentoring. This type of knowledge is shared by participants who show a researcher what they do, but either don't know what they "know" or couldn't express it in words. Likewise, this kind of knowledge can be shared with others via first-hand observation or videos.
Knowledge that *can* be expressed or articulated but hasn't yet been expressed, variously called *tacit*, *implicit*, or *explicit knowledge*.	It can be expressed and captured via interviews, conversation, or retrospectives, for example. Most obviously, participants share this kind of knowledge in research interviews, but researchers might share this kind of knowledge with stakeholders during conversations, collaboration, or share backs.
Knowledge that's been expressed or articulated but hasn't yet been documented, variously called *tacit*, *implicit*, *explicit*, or *verbal knowledge*.	It can be shared via communities of practice, shadowing, observing, or teaching.
Knowledge that's been documented, which is also called *explicit knowledge* or *information*.	It can be shared via any form of media: documents, slide decks, or audio-visual files, for instance. In the research context, this includes research reports, or any other media used to communicate research findings.
Knowledge that's encoded within the culture, strategies, and structures of the organization as per Professor CW Choo. Milton says: "Our organizations often operate the way they do as a record of past learning, either conscious or unconscious. That knowledge is nowhere written down but permeates the way the organization acts and responds."	It can be shifted and shaped by adjusting culture, strategies, and structures—and via operations. Operations are *instrumental* in shifting the research culture of an organization by way of the tools and training you offer, the people you hire, and the engagement and prioritization protocols that you offer. Or not.
Knowledge that's encoded into technology, or *machine knowledge*.	It can be embedded into tooling micro-copy, templates, mechanisms, or algorithms. As you will learn in Chapter 8, "Tactical Tooling," the tools you provide—or don't provide—and how you provide them are instrumental in shaping cultural knowledge.

THE BREADMAKING "TWIST-STRETCH"

Professors Ikujiro Nonaka and Hirotaka Takeuchi of the SECI model used this fascinating example[22] of how tacit knowledge can be transferred to machine knowledge. A team working at the Japanese company, Matsushita Electric (now Panasonic), was prototyping a bread-making machine, but they were having problems with dough consistency. To resolve the problem, engineers spent time working alongside master bakers and observed the unconscious "twist-stretch" movement of their hands. This bread-making know-how (tacit knowledge) was encoded into the machine's kneading mechanism (machine knowledge), and, once they'd fixed a few other quirks, the machine was a significant success. Academic debates[23] exist even about the absolute relevance of this story! But it's still a good story.

Defining an RKM Strategy

Your head may be spinning with theories and models for thinking about knowledge management in research. Now the question is: How can you put all this theory into practice? The RKM Model© (see Table 6.2 on pages 144–145) combines these theories and is designed to help you create a custom-fit strategy for research knowledge management. It works on the premise that knowledge management in research involves two distinct phases, each wholly different in intent (and vibe) from the other, and each in need of distinct tactics. The two phases are *communities of practice* and *knowledge services*, and a deeper delve into each follows. The RKM Model is a framework for addressing the full stretch of RKM needs across the research workflow. Use it to guide your choices and your conversations as to what RKM is all about.

---

22 Ikujiro Nonaka, "The Knowledge-Creating Company," *Harvard Business Review,* July–August 2007, www.hbr.org/2007/07/the-knowledge-creating-company

23 Rodrigo Ribeiro and Harry Collins, "The Bread-Making Machine: Tacit Knowledge and Two Types of Action," *Organization Studies* 28, no. 9 (2007), https://journals.sagepub.com/doi/10.1177/0170840607082228

Communities of Practice

Communities of practice is a term that has entered the business vernacular in recent years to describe self-made and managed communities of people who are interested in learning about the same thing, and they are a useful concept to harness in honor of research. Per the HBR article, "Communities of Practice: The Organizational Frontier,"[24] "The organic, spontaneous, and informal nature of communities of practice makes them resistant to supervision and interference." But operations can "provide an infrastructure in which communities can thrive."

Communities of practice express the creativity, spontaneity, messiness, and discontinuity of the SECI model's socialization and externalization. It's the phase in which researchers and stakeholders gather in the short- or medium-term to scope and plan a study, learn by listening and watching, capture countless messy notes, and create new (explicit) knowledge by analyzing, synthesizing, and documenting *together*. The social aspect of this phase means that creation and curatorial powers need to be in the creator's hands. Instead of centralized rigor and taxonomic control, ResearchOps should provide well-maintained knowledge "campfires"[25] that researchers can use to set up instant and customized social gatherings.

Campgrounds aren't permanent abodes, and gatherings are social, circular, generative, and unbound—like a music festival. Similarly, communities of practice are about *decentralized* control, of folksonomy, and easy-to-use tools that support people in collectively making meaning in their own way and time.

RECOMMENDED READING
SITUATED LEARNING

For a more intensive and structured view on communities of practice, see Jean Lave and Etienne Wenger's 1991 book *Situated Learning: Legitimate Peripheral Participation*,[26] published by Cambridge University Press.

---

24 Etienne C. Wenger and William M. Snyder, "Communities of Practice: The Organizational Frontier," *Harvard Business Review*, January–February 2000, www.hbr.org/2000/01/communities-of-practice-the-organizational-frontier

25 In 2014, while contracting with the UK's Government Digital Service, I wrote a blog post called "Vertical Campfires: Our User Research Walls," *Government Digital Service* (blog), September 3, 2014, https://gds.blog.gov.uk/2014/09/03/vertical-campfires-our-user-research-walls/

26 Jean Lave and Etienne Wenger, *Situated Learning: Legitimate Peripheral Participation* (Cambridge, UK: Cambridge University Press, 1991).

TABLE 6.2 THE RKM MODEL

SECI	Culture	
	Does your organization's culture support research knowledge sharing and use?	
	Socialization (Tacit to Tacit)	Externalization (Tacit to Explicit)
Research workflow	Scoping and planning	Collating data
	Setup	Analyzing
	Doing	Synthesizing
	Listening and observing	Documenting
	Sharing via conversation or research presentations	Sharing by involvement in knowledge creation

Though these phases are distinct, they should operate in tandem with one another and share assets and information with one another.

Culture or "vibe"	Communities of Practice
	Tight-knit and self-managed
	Created and managed by individuals and small teams
	Decentralized structure
	Folksonomy
	Usually short- or medium-term use
	No required submission criteria
	Services focus on supporting socialization and externalization
	Conversations, observations, raw data, nuggets, notes, tags, thoughts
	Assets may be abandoned when researchers and stakeholders move on

Tactics	ResearchOps can provide:	ResearchOps can provide:
	• Pathways to observe research and take part in planning, analysis, and synthesis sessions.	• Data management tools and protocols.
	• Literature reviews or secondary research to support scoping and planning.	• Analysis and synthesis spaces and tools.
	• Training and the means to do a literature review or secondary research.	• Training, protocols, and templates for creating temporary data repositories and folksonomies.
	• Access to relevant raw data stored in the repository.	• A protocol for moving a project from in-progress to closed (data should move into a centrally managed repository for future long-term use).
	• Protocol for moving data between "campgrounds," the repository, and the library.	• Data cleansing protocols.
		• Artificial intelligence, which may be useful in supporting coherence or search functionality.

Combination (Explicit to Explicit)	Internalization (Explicit to Tacit)
Accessing existing data or reports	Assimilating research content
Creating or consuming literature reviews or secondary research	Making decisions
Sharing via research presentations	

Knowledge Services

Centrally managed

Created and managed by knowledge management specialist/s

Structured taxonomy

Long-term findability

Submission criteria to manage quality

Services focus on providing access to research assets

Research reports, literature reviews, recordings of research presentations

Assets are centrally managed for long-term findability and relevance

ResearchOps can provide:	ResearchOps can:
• A self-service library and library service.	• Provide training on how to assimilate research content and how to validate it for relevance.
• A research repository with access by request-and-response.	• Enable stakeholders or researchers to log decisions made against research content, at least where the connection is still obvious.
• A literature review and secondary research service—both a self-service and full-service offering.	
• Training and guidance on how to access and do both.	

Knowledge Services

Knowledge services is the polar opposite of communities. Instead of a "bohemian" vibe—a research Woodstock—knowledge services thrive on control, i.e., strongly governed protocols and taxonomies. The goal is to carefully catalog and store high-quality research assets for future use by whoever might request it, which requires the provision of a service that can be accessed either on-demand or on-request. Here, ResearchOps will typically provide a library and a repository and their attendant services, such as a referencing service (a quick list of relevant resources) or literature reviews and secondary research in the case of a library.

Where the community phase is about fostering spaces for nonlinear knowledge creation, the focus of knowledge services is on *finding*. To make sure that finding is easy, you'll need to manage collections centrally and enable reliable and consistent search via a tightly controlled and structured taxonomy. You'll also need to make sure that both the search mechanisms and the content are trustworthy, which means that you'll need to set criteria and protocols for ensuring content quality like a review process or submission criteria. (It's likely easy to see why a skilled librarian is necessary.)

> **NOTE** LITERATURE REVIEWS VERSUS SECONDARY RESEARCH
>
> A *literature review* is an analysis of existing research on a topic, but without analyzing and synthesizing it for new insights. A librarian could offer a literature review to help researchers better understand what's already known about the topic before doing either primary or secondary research. Of course, with a good self-service library in place and with some training, a researcher could do this for themselves, too.
>
> *Secondary research* (or desk research) involves analyzing and synthesizing a range of existing research to gain new insights. Researchers can do secondary research, of course, but an appropriately skilled librarian could offer secondary research as a service, too.

Messiness and Trust

However well-laid your plans, to succeed at knowledge management, you'll need to address the cultural and emotive needs of the
organization and its people. More pointedly, you'll need to address
the tendency for humans to be messy en masse, and secondly, the
necessity of trust. Messiness needn't be handled in communities
of practice—let people create their own environments—but it must
be tightly controlled in knowledge services. This can be done by
governing the quality of inputs and instituting a structured and
well-maintained taxonomy, tasks that sit in the wheelhouse of a good
librarian. But trust needs more sensitive attention. There are several
layers of trust that must be navigated, and they are: trust in the
research organization, the individual researcher, the contents of the
research assets and insights, and the system that retains the research
assets (i.e., the library).

Because knowledge is so intertwined with notions of trust, and
therefore the success of research, it's a topic that pops up regularly
in this book. At the end of the day, no matter how phenomenal the
research reports are—reliable, well-timed, contextual, and useful—
how searchable the research library is, or how skilled your librarian
might be, if the team or person who produced the research isn't
trusted, all is for naught. While it is not incumbent on ResearchOps
alone to fix institutional issues of trust, there are things that ops can
do to support a positive cultural shift. Operations can:

- Work with research leadership to devise strategies to reform the
  culture, if needed. Trust can be a touchy subject but if it's impact-
  ing your efforts, you must broach it head on.

- Ensure that research is findable.

- Make sure that the services are well used (and used well) via
  onboarding and support, and monitoring feedback and metrics.
  A stagnated system that's not being used will erode trust (apart
  from being a waste of time and effort).

- Bring rigor to quality control via templates, training, and a
  governance protocol.

CONTROLLING CONTENT QUALITY

The primary goal of research is to give the right people the right information at the right time and in the right format, so that they can confidently make the right decisions. To succeed, your knowledge management efforts should meet the following criteria:

- **Right:** Knowledge should be accurate, reliable, and applicable in helping answer a particular question.

- **Informational:** It should enable someone to extract insight or new knowledge from it.

- **Accessible:** It should be easily available to the right people (the right permissions should be set, the right people should know where to find it, and it should be in the right place). Also, make sure that tools are accessible to people living with disabilities—a good rule for all operations.

- **Place:** It should be available in the right place so that it's easy to know about and access.

- **Timing:** As well as place, it must be available at the right time. Find-ability engenders trust in the completeness and competence of the source and means that people will return to the source of truth repeatedly.

- **Format:** It should fit the purpose and culture of consumption and communication and support knowledge exchange.

- **Consistent:** A regular and reliable format will allow people to learn how to consume research. For example, the scientific world has a typical format that includes a title, abstract, synopsis, introduction, materials and methods, results, discussion, conclusion, references, acknowledgments, and appendices.

- **Decisions:** It should support the making of decisions—the ultimate goal. Decision-making will usually result in an action and, therefore, "impact" in user research parlance.

Some organizations have used lists like the above to create a quality checklist for library submissions.

In a Nutshell

Knowledge is *the* primary product of research, so it's surprising that user research practices don't invest in knowledge management as standard. What other manufacturer doesn't heavily invest in its product distribution, store front, or web store? If you get the investment of an ops head count, place all bets on building operations that support knowledge creation and storage, and leverage and hire people who understand the craft. When that moment comes, use this manifesto to remind you of the backbone of this work:

- **Whatever the promise or the tagline, a tool alone will not deliver well-managed knowledge.** Instead, devise an RKM strategy and, again, hire the right people to deliver it all.

- **When you devise a strategy, consider how, and if, you'll address communities of practice and knowledge services.** They each need a different approach.

- **Knowledge management is about culture, community, and cognition, not just cataloging information, although that's a huge and ongoing task in and of itself.** So, don't forget to address communications and change management.

- **Efforts will need long-term, constant, and skillful iteration and attention.** RKM isn't finished just because a "solution" is delivered.

- **Calculate the accumulation of knowledge over time so that you can sustainably scale your RKM ops.** Stuff accumulates quickly, so make sure it's valuable.

- **Definitions and clarity around key KM words are important, but don't get bogged down in philosophy.** (Don't make them up either.) In the philosophically heady world of knowledge management, pragmatism wins all.

- **Lean into long-established professions and hire specialists from information science, librarianship, or knowledge management.** There's no need to make things up.

CHAPTER 7

Seamless Support

I've worked in organizations where researchers have been incredibly independent: they've thrived with the addition of fundamental operations like a research lab and a suite of fit-for-purpose tools, and they've needed little help in navigating how to use them. Point them in the right direction, and they've simply figured things out. I've also worked in organizations where researchers and, crucially, PWDR, or "people who do research,"[1] have needed a great deal of support to find and use things. In addition to guidance about tools, they've needed help with ethics and legal requirements, where to find a tricky participant cohort, how much to offer them as a thank-you gift, the best research methodology to use, and more. And if your operations are mature enough to include a research library, people will undoubtedly want help answering their questions or finding reports, too. The difference between the two scenarios isn't the intelligence of the people requesting support. Instead, it's that a tight-knit community of experienced researchers with limited operations in place can often help themselves quite well, but if you're operationalizing for a diverse and scaled-up (or scaling) research practice, you'll invariably need to do much more.

To successfully enable research that's scalable, you'll need to empower dozens, even hundreds, of people to navigate and operate an often-disjointed list of vendors, services, and tools all with different access requirements, user interfaces, relationships, and rules. Even a minimalist research workflow can feature a dozen or more discrete things to use making research feel more like a game of Snakes and Ladders—one hundred squares full of traps and tricks!—than a seamless system for research (see Figure 7.1).

---

1   *PWDR* is a term that I coined sometime in 2019 to talk about the cohort of people who do research as part of their role—designers, product managers, content strategists, marketers, and more—but who are not full-time research professionals. Since then, the term has become widely adopted.

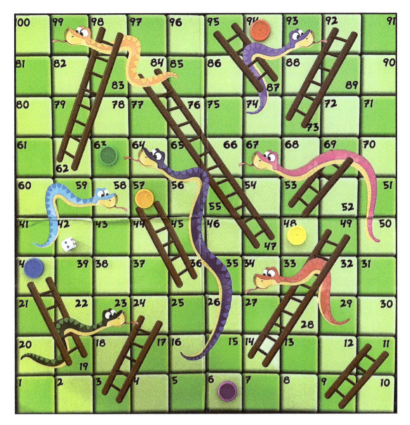

FIGURE 7.1

The children's board game, Snakes and Ladders, is a great illustration of how disconnected research operations systems experiences can be if they aren't thoughtfully designed to include onboarding and support.

More than simply being über helpful, publishing a handbook, or diligently responding to direct messages (DMs) and emails, the latter being hugely unscalable, you'll need to treat support as a system in its own right: one that's strategic, cohesive, and well-mapped-out. For without intentional design, support can quickly become unnavigable, unmanageable, and a fast trap to operational decline. But give support the attention it demands, and you can transform a disparate list of assets that is overwhelming into a seamless and empowering system for doing, taking part in, consuming, and administering research.

Maps and civic systems are a common theme in this book because they provide an abundance of examples for how to simplify sociotechnical systems, and it's no different when it comes to onboarding and support.

Be a Park Ranger: Give People a Map

If you've ever traveled to a national park like Yosemite or Yellowstone in the U.S., it's likely that you easily found a map to help you plan and navigate your journey (see Figure 7.2).[2] These maps tend to show a birds-eye view of the best hikes and views, and they offer practical advice, like where to find campsites, toilets, or food. Once you've arrived, the signage in the park tends to match the map precisely, and, as you get closer to your destination, it offers more detail, such as the exact distance to the campgrounds or historical facts and rules—no fires in June!

When you're knee deep in the day-to-day noise of operations, it can be easy to forget that everything that you provide, and every group of people that you provide it to, will create the need for a logical (and logistical) network of pathways, signage, and guidance. The knee-jerk reaction is to provide a list of links and resources on an intranet, or to fire up a new messaging channel—I've done it myself—but scaled-up research environments are often too disconnected, complicated, and variable for that kind of design trope, or instant reaction, to work.

---

2 "Yellowstone, Maps," National Park Service, www.nps.gov/yell/planyourvisit/maps.htm

FIGURE 7.2
You can take the idea of a map and other features of a national park, like park wardens and information kiosks, literally—it's visual and fun!

In the past, I've hired a service designer to take a step back: to map the most common journeys based on top tasks[3] and support requests, and to come up with a system for navigation and signage. If a service designer is out of scope, it's productive work to do yourself. Just remember that a map and signage (as pictograms or words) is only as good as it is consistent across the entire research operating system.

> **NOTE** **THE SLIPPERY SLOPE OF UNPLANNED SUPPORT**
>
> Most ResearchOps teams have minimal resources available to handle the day-to-day needs of support plus the work of building operating systems, which can be exhausting and eventually disabling. It's an easy mistake not to plan ahead for support when you're in deliver-the-thing mode or when you're still small, but the long-term success of your operations depends on this planning. It's why the ResearchOps Planning Matrix includes a section for onboarding and support and KTLO, which involves the ongoing maintenance, admin, and financial requirements of delivering *support* (see Chapter 4, "Planning Realistic ResearchOps).

---

3 Gerry McGovern, *Top Tasks: A How-to Guide* (Ireland: Silver Beach Publishing, 2018).

There are two types of support that operations can deliver to help an organization to scale research: logistical support (*how-to support*) and craft support (*what-to-do support*). Logistical support involves "How do I access XYZ or use this tool?" kinds of questions, whereas craft support involves "What methodology should I use for this study?" or "Where should I find XYZ participants?" These types of support require quite different approaches. For example, although logistical support isn't an option, it has a key advantage: it's much easier to scale.

Logistical Support: A Necessity, Not an Option

Every time you provide a system of any kind to a group of people, you'll need to provide some kind of how-to support—it is not an option. But you do have a significant degree of choice about *how* you provide support, i.e., the operating model you choose, which will predicate people's expectations, the drain that delivering support places on operations, and, crucially, the perceived value of operations. If logistical support isn't properly designed, resourced, and systemized, it can:

- Degrade researchers' experience of your operations.

- Precipitate dropout or *churn*: the rate at which people stop using your service within a timeframe, to use a product management word.

- Impact research quality and timelines. The goal of operations is to enable efficiency in research, not to slow or frustrate.

- Degrade operational buy-in...or upgrade it! But only if the cause is a lack of resourcing, and you can articulate it.

- Absorb the bulk of your time, which is fine if that's the intent of your design or where you can add maximum value. But administering support full-time rarely makes for impactful or scalable (never mind fun!) operations.

The instructional, rule-based nature of logistical support, or *explicit knowledge*,[4] means that it can be more easily documented, standardized, and automated; all ideal features of a self-service or light-touch

---

4   Explicit knowledge is often rule, fact, or process-based, and it can be easily codified, documented, and shared. For more about explicit knowledge, see Chapter 6, "Long Live Research Knowledge."

model. To support how-to questions, you might provide things like a research handbook, templates, automated emails, or pop-up tooltips (see "Choosing the Right Channels"). Also, answering logistical questions doesn't require expertise in research craft. For this reason, regardless of their research experience, ResearchOps professionals can handle the day-to-day admin, provided their time is well-managed and the support system is intentional. Finally, logistical support can be more easily outsourced to nonresearch professionals, say the workplace technology department if the need is tooling access, which can be a huge help.

Craft Support: Tricky but Scalable

Doing research involves a skillful and systematic approach to creating something new—in this case, knowledge. It requires careful planning, attention to detail, the use of specialized tools and techniques, and it requires practice and experience. It is *tacit*[5] knowledge, which is not easily codified, say in writing or videos, and it is context dependent. Researchers, particularly those who are up-and-coming, may on occasion want craft advice about a research study they're working on, say "What's the best research methodology to use for this study?" Or perhaps they want a second set of eyes on their discussion guide. In most cases, researchers have a strong research community to lean into, both within and outside of their immediate environment. But the context of democratized research can make craft support trickier: it's not uncommon for someone who's doing research as part of their job

---

5 Tacit knowledge is "learning by doing" or "know-how," and it's difficult to codify or express. The opposite is explicit knowledge. For more information about tacit knowledge, see Chapter 6.

to seek advice, a green light, or a confidence boost about an upcoming research study. To satiate this need, they'll often turn to their nearest and friendliest researcher, who may offer a great impromptu support "service." But good advice invariably generates the need for more good advice, which, over time, can place a significant and often unseen drain on researchers' time and energy.

If you work in an environment in which research activities are shared across disciplines, this scenario is likely familiar. If it's not seen as a problem, don't fix it—there are likely a million other things you could be doing. But if this dynamic is a drag on researchers, you'd do well to come up with a sustainable system that supports researchers in saying "no" when they need to and have the ability to pass requestors on to a formal craft support service.

Limit the Supply, Control the Demand

Requests for craft support tend to involve one-on-one interactions with someone knowledgeable about research craft—a researcher, usually—and about something highly specific. This means that craft support almost always needs to be delivered as a full-service model, which is trickier to scale. To offer craft support, you'll need experienced researchers' time—only experts can offer expert advice—who can monitor and triage questions, gain context, communicate, think through a study, and share their expertise. This means that you'll need an adequate supply of researchers' time to meet the demand, along with a way to *control* the demand. Again, great advice will tend to create the demand for more great advice, and sometimes an over-reliance on it, so you should be clear about why you are offering the support in the first place. By not having a purpose or a logical model, organizations often pinball from one craft support approach to another (coaching, training, mentorship, ad hoc advice, and more), but the service is quickly overwhelmed, and it's frustrating for researchers.

If you decide to deliver a craft support system, you must devise an operating strategy that addresses the mechanisms of supply and demand and prioritizes requests so that the service is as impactful as possible. For example, you might limit craft support to a particular set of teams, disciplines, or stakeholders, or a particular type of research. That is to say: where it will add the most value, and where there is financial investment to support it.

CRAFT SUPPORT: A PRACTICAL OPERATING MODEL

Because craft support is high cost, it is rarely a good idea to operate on a first-come-first-serve basis. You must always set some sort of limit. If you keep this rule in mind, there are practical ways to make craft support more sustainable and scalable. Here's a model that can work well in a democratized research environment:

1. Develop a foundational research training program and encourage anyone who asks for advice to attend it. This will help set a foundation of knowledge on which ad hoc advisors, whether researchers or a dedicated research coach, can build.

2. Limit one-on-one advice to those who have completed the training or parts of it. This doesn't mean that researchers won't ever be asked for advice, but by setting boundaries, you can empower them to say "no," or to time-box advice. Besides, it's a nice way to reward those who have invested energy in completing research training.

3. Finally, if resources allow, you might hire someone into the position of research educator or coach: a senior researcher dedicated to offering training, producing, and maintaining knowledge assets, and handling ad hoc advice. It's a unique and interesting position for a researcher who loves sharing their craft.

People who haven't done research training will inevitably have craft questions, too. Although they shouldn't be able to access one-on-one advice, ignoring them is out of the question. In this case, use timeboxed tactics to give them help. You could do the following:

1. Set up or use an existing group channel, perhaps in Slack or Microsoft Teams, to handle craft questions and then empower every full-time researcher, whatever their experience, to field and answer them. For simplicity, the channel could handle craft *and* logistical support. The fewer channels, the better.

2. Empower researchers to encourage people looking for advice to do research training. You might provide them with an easy-to-recall training sign-up link to reference.

3. Alternatively, or additionally, schedule regular craft office hours: a one-hour slot two or three times a week that anyone can book on a first-come-first-serve basis. You could rotate this among experienced researchers to share the load.

4. Publish a playbook and templates that document the research approach. Make sure that researchers have bought into the guidance, and that it's consistent with the research training.

Choosing the Right Channels

These days, even if you work in the same building as your colleagues, it's likely that the bulk of your communications and knowledge sharing happens digitally, whether via an intranet, messaging tool, or that most old-school of digital missives: email. It's worth remembering, that excellent support hinges on excellent knowledge sharing, which requires excellent communications. Outside of producing good content, the communications channels you choose to use will play a significant role in the success, or failure, of your support services—not just how successful you are at helping people, but how sustainable your operations are. There are plenty of options with several pros and cons, and they are the following:

- Onboarding
- Published support
- Help desks
- Messaging tools
- Office hours
- Coaching

Onboarding

Every system and subsystem that you deliver should include some form of onboarding: welcome—this is who we are, what we offer, and what to do next. Try not to see onboarding as a one-off event in which someone should learn everything there is to know about research. Instead, break onboarding up into manageable chunks so that you can deliver information when it's needed, and in context.

Say a designer who will do research as part of their role joins the company. The design team's onboarding includes a link to a two-minute, self-paced introduction about doing research, the result of a useful collaboration. The introduction gives the newbie a heads-up as to who you are, what you offer, and a memorable signpost for where to find out more when needed. New staff members are typically in a state of information overload, so the introduction is purposefully short and simple. As the designer settles into their role and they start doing research, they're provided with as-short-as-possible introductions to succeeding at the task they're facing, *when* it comes up, like recruiting participants, informed consent, or sending a thank-you gift. So, their knowledge grows over time.

This example assumes that onboarding is "chunked" and self-paced, which is a super-scalable model, and it also assumes that onboarding won't bottleneck anybody. You could offer research onboarding in a group setting or one-on-one, if a high-touch approach is more appropriate. If you opt for chunked and self-paced, you can also use the content to respond to ad hoc support requests, which can save time and build a culture of self-reliance while still being helpful.

Published Support

Publishing research guidance is an essential support feature in any scaling research organization. Published content could be written, shared as videos, or delivered as handy tools like templates that are, ideally, published in an easy-to-find and accessible space, like a commonly used intranet. You could produce and publish:

- Playbooks, which tend to document how to get research done from a methodological point of view, i.e., craft support.
- Handbooks, or how-to guides, which cover the how/when/where/who of common logistical tasks, i.e., logistical support.
- Toolkits, which help people easily find, access, and use a set of tools. Tools needn't just be technology, they can also include checklists, research templates, field kits, and even a map for the stationery cupboard. This is also logistical support.

It's useful to involve an information architect, knowledge manager, content strategist, or even an education designer, in your initial design and setup; support is all about communications, so kicking this work off with an expert is a smart place to start. (If your team includes a librarian, you might use their skills here.) I've had a full-time education designer on my team, and I would hire that role again in a heartbeat. There is always content to create and update and courses to make and administer, and the professionalism and ease-of-use they bring is a boon for the reputation of research.

THE DIÁTAXIS FRAMEWORK

There are fields of professionals dedicated to the art of educational documentation. Daniele Procida, Director of Engineering at Canonical, is a thought leader in the field of technical documentation, and he developed the Diátaxis framework. Diátaxis is a widely adopted documentation authoring framework that identifies four discrete modes of documentation: tutorials, how-to guides, reference, and explanation.

Tutorials guide the reader through a series of steps to complete a project or task of some kind. Tutorials are learning-oriented. In ResearchOps, this might be a tutorial about how to recruit participants for unmoderated research.

How-to guides are "directions that take the reader through the steps required to solve a real-world problem. How-to guides are goal oriented." To build on the previous example, a how-to guide might focus on how to set up a recruitment project in the participant recruitment platform.

References are technical descriptions of some kind of process, technology, or tool, and how to operate it. It's information oriented. In this case, ResearchOps may provide workplace technology partners with a succinct runbook that they can reference to deliver support for research tools.

Explanation is "discussion that clarifies and illuminates a particular topic. Explanation is understanding-oriented." You may use explanation to enrich your support content with real-world and contextual stories to help people understand *why* something is best done one way and not another.

A good support service will utilize all four modes of documentation appropriate to the need. Procida's website offers great guidance around creating support content: https://diataxis.fr

Additionally, to learn more about delivering great documentation, there's a global community called *Write the Docs* that's focused on just that: www.writethedocs.org

Keep It Minimal

Creating content can feel wonderfully productive, but it's worth remembering that you don't need to (and shouldn't) document everything, or every detail about every task. Apart from being overwhelming for users, keeping masses of content up-to-date is exhausting—and nearly impossible. Instead, aim to do the following:

- Keep original content minimal and, where possible, rely on signposts to content that others keep up-to-date, like support content that's published by vendors on their websites.

- Where possible, avoid using names of teams, products (other than vendor names, like User Interviews or Optimal Workshop), or people, because they're bound to change at some point, too. If there is a change, lean into "search and change" to make amendments en masse.

- Before putting "pen to paper," work out an information architecture. Do a Top Tasks study (see "Define Top, Difficult, and Important Tasks), or something similar, to document what people genuinely need to know about, and then let everything else be answered by social exchange, i.e., on Slack or Teams, or in the hallway.

- Lean on formats like checklists, task-based titles, and a workflow-driven architecture to make content easy on the brain and easy to navigate. You might pick up a copy of Atul Gawande's book, *The Checklist Manifesto: How to Get Things Right*.

- Keep your language simple and straightforward. You're not looking to gain a Pulitzer in writing! In fact, overly expressive and detailed content tends only to result in overwhelm for your audience.

- Try to maintain a similar structure, format, and tone throughout your documentation. All your support content, whether written or video, should feel like it comes from the same team; otherwise, it may not feel accurate or valid.

- Deliver content via various forms of media to meet the needs of different needs and learning styles. As per the Diátaxis framework, tutorials (learning-oriented), how-to guides (task-orientated), technical references (information-oriented), and explanations (understanding-oriented) may need a different approach.

- It should go without saying, but make sure that all of your content and the tools that you use are accessible by people living with disabilities.
- Finally, do a quarterly or annual audit of both content quality and metrics as part of your regular operating reports. (For more on operating reports, see Chapter 10, "Money and Metrics.") You should measure things like most commonly searched terms, content that's most and least used, and any other feedback, and then shift operations in response. Do a content audit to archive out-of-date content too. The tidier your house, the easier things will be to navigate.

A BONA FIDE SOURCE OF TRUTH

Keeping content minimal and to-the-point solves for another common and maddening trip-up of published support: maintaining a source of truth when duplication is rife.

You might work tirelessly to publish and maintain a beautiful handbook—a pinpoint accurate source of truth about doing research, and then people across the organization duplicate it and customize it to their own use. Even if pages are locked down, people tend to copy bits and pieces, edit them to suit their own context and style, and publish new versions that look bona fide. If this sounds familiar, take comfort that you're not alone. Of course, no one means harm, but when guidance changes, and it always does, disconnected and duplicated content won't keep up with the times. Even with the best content management technology on your side, it can be hard—impossible even—to maintain a single source of truth for support and guidance.

So, what should you do? It's unpleasant and unfriendly work to shut duplicated pages down because they represent someone else's unique needs and work. Besides, these pages often offer excellent feedback about how your content could be improved. I've seen visual designers make okay ResearchOps content look gorgeous! And they've inspired us to do better, too. Your content platform might enable you to manage duplication technically and, if so, by all means do it. If not, your best bet is to learn from duplications and steal their best bits. Build community and leverage duplicators' energy to strengthen your support content so that, ideally, they no longer feel the need to duplicate, because their voices are heard, and their needs met.

Help Desks

Whatever your scale and audience, a help desk is a sure-fire way to bring organization and efficiency to your support process. Help desks, also called *support* or *service desks*, are commonly used to manage support at scale, and they have many pros. A help desk like Zendesk, Atlassian's Jira Service Management, or Salesforce Service Cloud, will allow you to:

- Offer a seamless and routed request-and-response service for all kinds of support.
- Systematically handle a large volume of requests for support.
- Track and report support performance. This data can be used to make the volume and breadth of support more apparent; make sure to leverage it to pitch for additional funding.
- Support multichannel engagement. You can funnel requests for support via email, chat apps, and more into one support channel where it can be handled en masse.
- Attach your knowledge base to a support workflow.
- Enable all sorts of automations and integrations.
- Provide metrics for continuous learning and improvement.

All of this is excellent news for scaled-up support. Your organization will likely have a help desk tool already available and widely used, so you might not need to procure or budget for a tool either. If you choose to go the route of a help desk, and I recommend you do, set an SLA (*service-level agreement*) up-front, that communicates when someone should expect a response and a resolution to a question. Also, don't forget to build in protocols for using existing onboarding material, like snappy videos or links, to answer common questions (along with a friendly note).

Messaging Tools

Messaging tools or *chat apps* are ubiquitous to most modern organizations, the most common being Microsoft Teams and Slack. Most people's working lives, particularly as a result of the pandemic, center on some kind of messaging tool, and it's often the first place they turn to for help. For all the strengths that messaging tools have—a familiar platform, quick and open communications, shared learning, building of community—they have several downsides, too.

Putting DMs aside for now, that will be covered in a moment, group channels can:

- Make support requests hard to track, particularly if answering a question needs more than an instant response (unless you funnel messages through a help desk as previously noted).

- Be notoriously noisy and overwhelming, so questions are easily missed.

- Be a sinkhole for excellent knowledge; unless you make a concerted effort to capture it elsewhere or encourage habits and skills around using advanced-search features.

- Set the tone of an open-door policy, which is great in most contexts but less so in scaled-up support. It's impossible to be everyone's prompt and personal assistant when you need to support dozens or hundreds of people.

But you can get ahead of the downsides of group messaging apps, and here's how to do it:

- **Avoid setting up individual channels for different services or tools.** A single support channel should help *everyone* know where to get a question answered for *all* questions. On the other hand, it can be handy to set up a unique channel for a particular group, for instance, researchers versus people who do research.

- **Devise a channel strategy and put in some rules.** In other words, make sure that you know what your *standard operating procedure* (SoP) is for running a help channel: You'll want guides around topics such as who should respond, how long a first response should take (an SLA), how to make sure that a request is resolved, and what happens when it is. Your SoP can also help you understand your support workflow and where you can templatize messages or automate things.

- **On that note, automate actions where it will help.** You might enable channel admins to instantly turn a message into a help-desk ticket, which is useful for requests that will need a follow-up response. You could create a run sheet of common responses that can be quickly copied and pasted. Even better, you could code a handy text-expander tool, like textexpander.com, for super-fast templated responses.

- **Avoid giving custom answers to common questions and instead point people to existing published resources.** This will save time repeating information and encourage people to self-service in the future.

- **Where possible, distribute ownership of answering questions among a group—or decentralize support.** In a democratized environment, you might encourage a team of researchers to provide answers to questions in a group channel used by people who do research. Just make sure that they have access to the right information and can offer good advice.

Office Hours

Office hours are prescheduled hours that people can book on a first-come, first-served basis to access support. Perhaps ResearchOps offers the opportunity to book office hours to access participant recruitment advice, a common office hours topic, or any type of how-to advice every Tuesday and Thursday between 1–2 p.m. Prebooked time slots tend to work better than an open-door policy. It allows support providers to be as efficient as possible with their time and eliminates meaningless waiting around time. Also, make sure to request context for the support via the booking form so that providers can come prepared or even answer the question up-front if it's easier than it looks.

Office hours are an excellent adjunct to self-service support because the investment is timeboxed. They are a relatively small and controlled investment that can deliver high value, whatever your scale, and they offer an opportunity to connect with people about their feedback and needs firsthand, i.e., they're personal *and* manageable.

Coaching

Coaching is the highest touch (or full-service) of the support options, and it offers people who do research one-on-one sessions focused on building research craft and knowledge. Coaching sessions are an excellent adjunct to research training—people who have completed research training often need support as they apply their new skills in the real world—but they're an energy sinkhole if used for anything else, so make sure the purpose remains craft-focused. If enthusiastic people aren't supported in using new skills in the field, they may revert to past habits because they're under pressure to get a job done and old is familiar. Ideally, coaches will have a detailed understanding of the organization's ResearchOps resources to support continuity and consistency in advice.

Coaching sessions work similarly to office hours in that a research coach might offer a few hours a week that can be booked by support seekers, so that they're prepreprepared and timeboxed. While coaching sessions should be hosted by a professional researcher, it's incumbent on operations to provide the awareness campaigns, booking capabilities, attendance metrics, scheduling, and more, so that coaching sessions are successful (and tracked).

You Need a Support Strategy

The CEO of Apple, Tim Cook, is quoted as saying "We have one hardware organization. We have one software organization. It's not like we're this big company with all these divisions that are cranking out products. We're simpletons." If Cook can make a behemoth tech company structurally so simple, you can do the same for research. To be able to manage the behind-the-scenes complexity of scaled-up support, you'll need to design a system that makes the complex seem simple, while being sustainable and delivering business value. To do this, at the micro and macro levels, you'll need to:

- Understand the support context.
- Define top, difficult, and important tasks.
- Map your existing support journey.
- Define an operating model (and often several).

WHO TO HIRE TO HANDLE SUPPORT

As your operations move into the KTLO phase, you'll need to hire people to deliver support operations and the support itself. Ideally, you'll need:

- A senior leader and systems-thinker to design and deliver a support strategy and operating model and get buy-in for the investment needed.
- A content or education designer who can own the design and upkeep of varieties of onboarding and support content. You should at least work with a contractor to get this work done at the outset.
- At least one person whose primary focus is to deliver support to the people who need it. They should be in charge of the service desk, get back to messages, and make sure that office hours and coaching sessions run like clockwork.
- If you choose to offer craft coaching and training, you'll ideally need to hire a senior researcher with an interest in pedagogy. Researchers who have been hired to do research often don't have the time, skills, and sometimes the interest, to invest in this ongoing work.

Understand the Support Context

It's likely that your organization already offers support services to its employees, including you, and it's worth taking advantage of them. By aligning research support with existing support services, you'll ease the burden on your own operations and smooth the journey for end users as they move between contexts. (This point also applies to customer support, which implies support for research participants.) To deliver support that's well-integrated, ask yourself these questions:

- How have other parts of the business, say workplace technology or human resources, set up support? Are there opportunities to replicate excellent support practices?
- Who offers support within the organization, and could you partner with them? For instance, the team who runs the IT service desk might offer efficiency via collaboration or a complete handover of research technology support (a decentralized support model).

- Are there any tools, like a help desk tool, that are commonly used to enable support across the organization? If you find things that work well and are already widely known, reuse them. Scalable support is not the place to be unique.

- How do people learn or communicate in the organization? Is the emphasis on chat, written pages, images, GIFs, or video, perhaps? If there's a particular tone or style that works well and is recognizable, you would do well to replicate it.

- Will you need to offer support across multiple time zones or languages? How does the rest of the organization accommodate it?

- Crucially, does the research strategy, or any other strategy, signal the prioritization of support for one discipline, activity, or team? Perhaps a team of researchers are doing sensitive or high-priority research, and additional levels of support will add value.

Define Top, Difficult, and Important Tasks

To keep content minimal and put energy in the right place, you'll need to understand what people need help with the most, so you'll need to do research. I'm going to assume that you're surrounded by researchers, or you're a researcher yourself, so you've got plenty of know-how to lean into. You could use Gerry McGovern's simple but powerful Top Tasks methodology[6] to gain an understanding of top research tasks. By proxy, the information you gain will allow you to answer these important questions:

- Of the top tasks, what do people need support with the most? Could you reduce support requests by redesigning something, providing written guidance, or embedding a solution in context, say as a template or tip within a workflow?

- Are there infrequent tasks that are also important to succeed at? Infrequent tasks often need more intensive support, but if they're important to get right, give these tasks the attention they deserve. Perhaps a participant emails a researcher to request that their data be deleted. What should the researcher do?

---

6 McGovern, *Top Tasks*.

- What about difficult tasks, even if infrequent? Some tasks are just plain tricky and, if they're a blocker to progress if not overcome, may need intensive support, which can be a drain on everybody's energy. Perhaps you could alleviate this by ironing out the difficulty through better design or built-in instructions.

Use this information to create a prioritized list that defines who, and which tasks, are most to least important to support and, therefore, which channels to use.

Map Your Existing Support Journey

To form a well-prioritized support strategy, you'll need to understand the strong and weak points of the existing research experience. By creating journey maps[7] for prioritized people and tasks, you'll gain a solid understanding of the current experience you're delivering, whether intentional or not, and what needs to be done first, second, and third to deliver a more cohesive experience. Your support journey will be unique to your context, but whoever you support and whatever the format, each support journey will likely look something like this:

1. **Awareness:** How do people know that you exist and that you can help them?
2. **Onboarding:** How well do you give people the basic access, permissions, skills, and knowledge they need to get going?
3. **First-flight jitters:** How successful are people in running their first, second, or third research study when their knowledge is still new? At least in the case of researchers or people who do research. Don't forget about other people, like research consumers who might need help doing their first search of the library.
4. **Ongoing support:** Once someone's confident to go it alone, how well do you offer ongoing support for ad hoc questions and problems?
5. **Change management:** How well do you consult, inform, and support people when there's a change in what's available or how things work?
6. **Off boarding:** How well do you manage research data and accounts when someone moves teams or leaves the organization?

---

7   A journey map is a common service design asset. It's a visual representation of the steps a person goes through when engaging with a product, service, or experience.

These "six stages of support" should form the horizontal plane of your support journey maps. The various channels you use to deliver support will create the vertical plane.

RECOMMENDED READING
CREATING JOURNEY MAPS

Journeys maps are used to describe step-by-step how a user interacts with a service and their pain points and emotions at each stage. How you use a journey map is up to you, but here's a standard template by Service Design Tools:
https://servicedesigntools.org/tools/journey-map

Define an Operating Model

To develop a support operating model that scales well, you'll need to make smart choices about the channels you choose and the tactics you use to create your *front-of-house* and *back-of-house experiences*, both handy service design terms. You'll likely use several different tactics across your support experience: some contexts may need full-service decentralized support, on-demand, or request-and-response, while others may need self-service and centralized support, for instance. Each model has pros and cons, which you should weigh against your resources, strategic priorities, user needs, channels, and whether you're innervating logistical or craft support.

How you deliver support, or your support operating model, must sync closely with the operating model of the services you're support-ing. If you offer full-service participant recruitment, for instance, the level of support you'll need to offer will be minimal—you're doing the bulk of the work, after all. But if you give people the resources that they need to do recruitment on their own, depending on how well-designed your operations are and how skilled and confident your users might be, they may need a good deal of support.

The following operating tactics were covered in Chapter 3, "From Strategy to Operational," but they're worth reviewing in this context:

Self-Service Support

Self-service support focuses on helping people help themselves. It's the most distant form of support and tends to involve on-demand or published content, tool tips and templates, self-paced training, and automation. If you've got the ability to automate answers to basic questions via a bot, by all means do so, but make sure that it works right. When an automation goes wrong, it can cause more frustration (and therefore more support work) than simply offering human help at the outset. You don't want to create more noise than value.

Full-Service Support

Full-service support means that you offer one-on-one attention to help someone get something done, without actually doing it for them. Note that full-service support is distinct from a full-service offering, say participant recruitment, which involves achieving the actual outcome for someone. This form of support usually involves a request-and-response model. You might enable support seekers to submit a help desk ticket for one-on-one help, provide office hours, respond to messages in a group channel, which has the benefit of making support open to view for all, or accept meeting requests, emails, or direct messages. Meetings, emails, and DMs have significant drawbacks, as covered in this chapter under, "Choosing the Right Channels."

Centralized Support

Centralized support means that ResearchOps is the single point of contact for support. In other words, whether the support is self-service, full-service, or on-demand, etc., research operations will manage the support experience. (The opposite is decentralized support, which is covered next.) The advantage of centralizing support is that you retain full view of the themes emerging from support, which means that you can resolve common issues through service iterations or a full redesign, thereby reducing support.

But the disadvantage is too great to ignore: it's challenging to gain head count for unglamorous work like support, even if it's vital. As a result, centralized support can be hard to scale, particularly when you're still growing your impact. And if the people or person who handles support goes on leave (or leaves entirely), support services can grind to a halt, compromising the most important characteristic of good support—continuity. In short, if you're a small and maturing operations team, it rarely pays to centralize support entirely. Instead, decentralize support by looking for partners to take on some of the load.

Decentralized Support

Decentralized, outsourced, or distributed support means that you've enabled other people to help you deliver support. It can be helpful to sacrifice centralized control for all the help you can get, but there's another advantage, too: it's quite often the case that other people or teams will be nearer to the people looking for support than you. For instance, researchers are often the first person a research participant will turn to if they have a question about their thank-you gift or consent form. People who do research will often turn to workplace technology for access to a tool, or they may turn to a vendor for help using a platform. It's useful, therefore, to formally empower (or support) these first point-of-contact people to offer at least some level of help. You can alleviate the level of support you need to give these providers by handing over standardized tasks that are documented in handbooks, runbooks, or SoPs.

Decentralizing support does, however, bring a couple of drawbacks: You'll lose direct control and view of the bulk of support requests being handled, and people may customize support to suit their own context and view, which isn't all bad, but it does mean that you'll lose a single source of truth. Particularly if you're in the early stages of ResearchOps maturity, the loss of this crucial data can hamstring your ability to become an informant of research activities and strategy to leadership, which is a shame.

Finally, if you choose to outsource elements of your support journey, think about the impact on the end-to-end support journey. Researchers will often need to move between multiple vendors and services within the space of a single study—100 squares full of traps and tricks!—so it can be inefficient and confusing if they need to navigate a mishmash of partnerships that work for you but are discontinuous and confusing for them (all of which can result in more support work for you). Still, done well, the pros of decentralizing support outweigh the cons for sure.

MAKING INVISIBLE WORK VISIBLE

Support is big and important behind-the-scenes work, but it isn't sparkly or the next-big-thing, so it can be a challenge to gain buy-in. To build essential support for support, you'll need to continually make this *invisible* work *visible*. To do this, use visualization, metrics, and storytelling to make the extent of the work known. You might track the number of coaching hours offered and gather anecdotal feedback from the people who have benefited, including researchers who may be less distracted by ad-hoc requests for craft support.

Verbatim testimonials can be powerful. Make sure to track the number of support tickets you've resolved over the course of a month, quarter, or year, and make that number visible to your leaders. Centralize *all* support requests in a single channel, which will automatically produce a great visual asset that will bring your day-to-day support efforts into sharp view. Finally, use service design assets, such as a user journey or service blueprint, to help make the extent and complexity of the work visible. (See Chapter 10.)

In a Nutshell

You might have heard the quip that an excellent waiter is a waiter you don't see. Or perhaps you're familiar with this quote from the *Futurama* television series: "When you do things right, people won't be sure you've done anything at all."[8] The notion that you can design out the need for support is a fallacy; however excellent your systems are, people will *always* need some level of support. But you can design support so that it is vastly less onerous to deliver, and so that it helps people navigate, learn, and succeed in increments that are low effort. To do this well, you'll need to:

- **Design support as if it's a journey, not a multitude of resources and tools.** The notion of navigation is a frequent theme in this book, and for good reason. Every time you add a resource or tool to your tooling stack, consider how people will know about it, navigate toward it, and what they'll need to do and know to achieve their goals.

- **Make choices about the kinds of support that you'll offer.** Logistical support is not an option, but you should carefully consider if and how you'll deliver craft support—it may not be needed at all.

- **Understand the pros and cons of various types of support channels.** When you design a support operating model, be sure to make smart choices about how and when to use each channel, and how they should interact (or not) as part of a support journey. Used in the right place and way, automations and integrations are excellent to leverage.

- **Define a support strategy: a prioritized plan of what you will—and will *not*—support, and how.** No matter how much you spend or how hard you try, it's unlikely that you'll be able to support everyone all of the time. A support strategy will help you invest where support will be most valued, create a culture of self-reliance, and build the reputation of research where it matters most.

---

8 *Futurama*, IMDb, www.imdb.com/title/tt0149460/

CHAPTER 8

Tactical Tooling

In 2018, I published the "User Research Toolbox,"[1] a catalog of 274 researcher-recommended tools. I stopped adding to the list sometime in 2019—it became tedious to maintain—but if I'd continued to curate it, I've no doubt it would have quadrupled in size, at least. These days, for every research problem, opportunity, or methodology, there's a sharp and shiny purpose-built research tool, but it hasn't always been that way. Just a decade ago, if you wanted to create a research video library, you'd have turned to the tools of the broadcasting trade; to manage an in-house participant panel, you'd have looked to the worlds of sales and marketing. The maturing of the research profession means that small research technology companies—or "ResTech," for short—have become large, and startups are popping up everywhere.

The world is now awash with tools that promise to make research faster, more collaborative, and more impactful than ever before with bylines like: "Within a few hours, get the human insight you need to deliver exceptional products, services, or brands." "Collect actionable user insights that fuel product decisions in hours, not days." And even "User research without the users" compliments of AI.[2] It's bylines like these that are fueling the debate about what constitutes the tooling needs of the research profession versus tools that are designed to enrich the research technology trade. Do artificial intelligence, automation, and souped-up speed go hand-in-hand with reliable research? Whatever your take, as Chief Scientist for Software Engineering at IBM, Grady Booch, famously once said: "A fool with a tool is still a fool."[3] And so, it might follow that a fool with a fast tool is an even faster fool!

While culture is made manifest by many things—people, processes, and rituals, to name a few—it is also made manifest by tools: the tools you provide or don't provide, and how you provide them or to whom. These factors have enormous influence on how people do, perceive, and take part in research. The growing demand for research and the proliferation of research tools mean that you must, more than ever, be tactical (and tactful) so that tools not only scale

---

1   Kate Towsey, "User Research Tool Box," Airtable Universe, last updated May 26, 2018, www.airtable.com/universe/exp7BidtSB73ihAqw/user-research-tool-box

2   A product called *Synthetic Users* offers that you can "Run your user and market research with the most human-like AI participants."

3   Grady Booch is commonly quoted as being the first to say, "A fool with a tool is still a fool." But it may have been Ronald Weinstein, an eminent academic pathologist, who was the first to say the phrase, no later than 1989.

more research capability to more people, but scale a great research culture and practice, too.

Crafting Research Culture

In recent years, I managed ResearchOps for a company in which hundreds of people did research regularly. When I first arrived, there was just one research tool available: a much-loved, unmoderated research tool. If you were a designer or product manager keen to understand customers, "doing research" typically meant running an unmoderated test overnight and downloading the results the next day. Because the tool made research seem *so* quick and easy, it meant that fewer people spent less time engaging with real customers in real time, which degraded both customer empathy and an appreciation for well-executed research. In comparison to instant insights, researchers (and recruitment logistics) seemed slow, outdated, and pedantic. So why would anyone work with a dedicated researcher, or meet with a customer face-to-face?

The tool wasn't the problem, tools are rarely the full problem (or solution), but it exacerbated the problem. So, we asked: "How could we create a culture that appreciated well-executed research over 'research lite,' even if the research took significantly more time and effort?" As a research organization, we did several things to move the cultural needle. From a tooling perspective, we shut the unmoderated tool down and provided operations across the eight elements, including a minimal tool stack, to exclusively support moderated qualitative research, i.e., spending *real* time with *real* people. The strategy took boldness, time, and commitment to show results (two years, to be precise), but it helped forge a richer research culture and practice that paid back in dividends over time. As the saying goes: no guts, no glory! But also, no strategy, no designated outcome.

More than just siloed units of mechanical utility, tools create culture, and they shape craft—or its lack. By way of the tools that you provide—and *don't* provide—and how you stack them up, you can speed people up or slow them down, encourage them to skim the surface or dig into the details, incite close collaboration, or help people go solo. So, when you create a tooling strategy, consider the culture within which you operate and the culture that you want to maintain or create over time. Research tools should be guided by more than user requirements and features.

Sticky Notes and Sharpie Pens Are Also Tools

It's easy to become myopic about tools when there are so many flashy and excellent purpose-built tools from which to choose. But tools are simply instruments for getting things done and they come in all shapes and sizes. Excellent research tools needn't be digital, they needn't require complex integrations, and they needn't come with a hefty price tag, although many do. Take that much-loved and humble combo of the research wall, Sharpie pen and sticky notes. If you want to make friends with a researcher for life (and foster a hands-on research culture), provide an unlimited supply of these simple yet powerful tools. Mainstream tools can be bent to suit research, too. How about WhatsApp as a robust field communication tool and, if you're in a pinch, a diary study tool? Trello makes for a nifty short-term insight management tool,[4] though it doesn't scale well for that purpose. And Zoom works well enough for remote interviews. In short, think outside the box because these days you're spoilt for choice. Research tooling can include:

- Generic digital tools that also work well for research, like video conferencing tools, intranets, digital whiteboards, scheduling solutions, and egift card platforms that make sending participant thank-you gifts a breeze. You could also procure an established library management system (LMS) to deliver a robust research library.

---

4   Trello and similar tools can be used to support communities of practice or decentralized knowledge "campfires" per Chapter 6, "Long Live Research Knowledge."

- Specialized digital research tools that support various research methodologies, like diary studies or unmoderated research, logistics such as participant recruitment, or research knowledge and data management. Custom-made research templates, say for participant recruitment, can also be viewed as a tool.
- Handheld tools like video cameras, whiteboards, sticky notes, notepads, pens, cameras, mobile devices, audio recorders, and tablets can be useful. Then there's the odd handheld tool made especially for the research context, like design researcher Nick Bowmast's Mr. Tappy.[5]
- Physical spaces, like user research labs, observation rooms, interview pods, device labs, and research walls, are a type of research "tool" as well. The more technical of these spaces are often reliant on a sizeable and interconnected collection of tools.

More Than Just Procurement

It doesn't take long to learn that every tool you onboard into your workflow comes with a long list of tasks to keep it available, useful, and used (see Figure 8.1). Even tools like sticky notes and Sharpie pens, as simple as they are, require ongoing management of supplies, suppliers, finances, and more. And when a vendor adds or changes an interface or feature, as they often do, it can trigger a cascade of tasks to make sure that operations stay smooth. Yet mention the word "tooling" and most people's first response is: "Tell me how to speed up procurement! Why is it so slow? How long before I can call this job done?" But procurement is a fraction of the work.

In many established organizations, procurement and the requisite security and legal approvals involved in onboarding a tool can take at least three to six months. While the procurement process can be frustrating, if you view procurement accurately (as just one of *many* tasks involved in successfully setting up a tool), you'll find that the stop-start-drawn-outness that typifies procurement actually buys you time you need, sometimes desperately, not just to deliver a tool but to deliver it so that it adds maximum value.

---

5 "Mr Tappy," www.mrtappy.com/

Annually

Contract renewals.
Strategy review.

Quarterly

Planning the budget.
Reporting on metrics.
Optimizing in response to metrics.

Monthly

Vendor meetings.

Daily–Weekly

Administrating the tool.
Providing training and support.

FIGURE 8.1
Every tool that you onboard will require an annual cadence of near-constant work to keep it available, useful, and used.

Managing a couple of tools is one thing—one savvy human being can get that right—but it's not unusual for mature research teams to need a dozen or more interconnected tools, and the work can quickly add up. This investment means that tooling decisions can't be driven by short-lived wish lists and knee-jerk reactions or managed as a part-time job. Instead, decisions must be guided by committed strategies which should, in turn, help define a *research tooling strategy*. A tooling strategy will help you:

- Gain clarity on what you want to achieve both practically and culturally by way of the tools you provide—and *don't* provide.

- Know when to say "yes" and when to say "no." It's not possible or impactful to give researchers every tool that they want. Instead, you should provide tools that specifically innervate a strategic priority or satisfy a basic operational requirement, like an egift platform for participant thank-you gifts.

- Be minimalist, though not stingy. The right tools used right and managed right beat a plethora of ill-chosen and ill-maintained tools. Besides, most operations teams can't afford the inefficiency of excess tools, either financially or operationally, and poorly delivered tools (or too much choice) can result in cognitive noise for everyone.

- Your tooling strategy should include how you work with researchers who have "gone rogue" and started using an unapproved platform. This isn't a problem because ops are control freaks, but because organizations often have security and contractual requirements, and for good reason. To this end, you might offer a tooling backlog that researchers can communally contribute to. It's unlikely you'll resolve every ask, but at least people will feel heard, and you'll surface common requests.

- Make sure that the cost and effort involved in onboarding a tool is at least commensurate with its impact over time.

- Work with research leaders to plan ahead so that the readiness of the tool is well-timed. If a strategy is time-sensitive, and good strategies *always* are, you'll need to leave as little as possible to chance.

- A new tool isn't just a lift for operations. People who use tools will also need to adopt, adapt, and learn, which will inevitably slow them down and, in excess, cause frustration. Strategy will help you deliver tooling continuity and, where change is necessary, the bandwidth to manage it proficiently.

NOTE **IN THE FACE OF INEVITABLE CHANGE**

John Lennon's lyrics, "Life is what happens to you while you're busy making other plans"[6] says it all. Strategies evolve over time, they change shape, and sometimes they disappear entirely. I once onboarded a tool to support a benchmarking program, but not long after the tool was in situ, the organization shifted focus and the tool was deemed ill-suited for any other use. Even with strategy on your side, you won't get the tool or the timing right every time. Still, you'll win more often than not.

A Research Tooling Blueprint

User Interviews, a participant recruitment platform, has developed a reputation for producing some of the most entertaining, and useful, content that the field of user research has seen from incentive calculators to reports and tools maps. Their *2023 UX Research Tools Map*[7] does a great job of illustrating the expanding world of user research software (see Figure 8.2). The growth is astonishing. But the map also illustrates another important point: doing research is often a multi-tool journey.

FIGURE 8.2

A segment of User Interviews' *2023 UX Research Tools Map* shows just how complex a research tooling stack can get—and how much choice there is.

---

6 John Lennon, "Beautiful Boy (Darling Boy)," *Geffen*, April 11, 1981, www.azlyrics.com/lyrics/johnlennon/beautifulboydarlingboy.html

7 *The 2023 UX Research Tools Map*, User Interviews, www.userinterviews.com/ ux-research-tools-map-2023

Every time you onboard a tool into your research environment, you'll need to consider how people (and data) find, engage, and depart from a tooling experience, and where they go to next. It's not only researchers that you should think about; you must consider panelists, participants, consumers, stakeholders, administrators, and vendors interactions, too. Then you'll need to define how each tool should integrate and interoperate with other tools, people, practices, and data stores across the research workflow and, ideally, the workflows of adjacent disciplines, too. For instance, if researchers typically do *affinity sorting* using a wall and sticky notes and then need to share their analysis remotely, they'll need to be able to switch from a manual context to a digital one without too much of a fuss. If researchers need to work with designers or developers to ready prototypes for usability research, they'll need to move between various design and development systems to achieve their goals.

To understand this complexity, and design in relative simplicity, it's useful to adopt one of service design's primary templates, a service blueprint—or a *research tooling blueprint* (see Figure 8.3 on pages 186–187) to fit this context.

There are plenty of good reasons to map the variety of research workflows that you support, including to map support requirements, per Chapter 7, "Seamless Support." A research tooling blueprint begins with a specific research workflow, and then provides an aerial view of the following information:

- The phases that define that particular workflow.
- Individual tools needed at each phase of the journey to solve particular needs.
- Where and how various people, data, and technology will intersect—and *why*.
- The front-of-house tooling process: the tasks that your end users will do.
- The back-of-house tooling process: the tasks that operations administrators will do is important, too.
- Opportunities to consolidate the needs of multiple tools into single services to enable efficiency, cohesion, and simplicity.
- Opportunities or gaps in data interoperability that need to be filled, either by technology or manually.
- The full extent of the work required to achieve your tooling goals—or dreams!

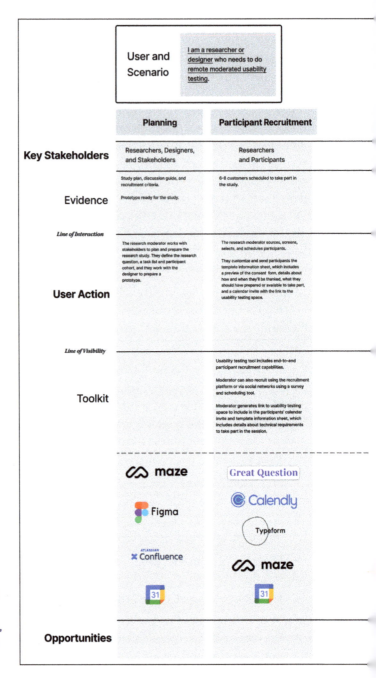

FIGURE 8.3
A research tooling blueprint that shows the high-level tooling needs of a research workflow. In this case, in-person moderated research.

	Set Up	Moderation	**Analysis** Thank Participants	**Deliver Findings** Close Study
	Researchers and Observers	Researchers, Observers, and Participants	Researchers, Collaborators, Participants, and Operations	Researchers, Stakeholders, and Operations
	Observers scheduled and briefed on the session. Signed observation etiquette forms.	Signed participant consent forms, video recordings from the testing sessions, and notes from the session.	Raw data and findings cataloged in the analysis tool. Insights as a result of successful analysis of all data. Participant has received branded egift card in their email inbox.	Research assets shared with stakeholders via the internal wiki, Slack, and share backs. Research report and share back recording are stored in the research library. Raw video data is stored in the research repository.
	The research moderator invites observers to the sessions, asks them to sign the observer etiquette form, and sets up a meeting to discuss the research objectives and how to take notes and communicate during the sessions.	The research moderator: • Notes the NDA and lets the participant know that the prototype should not be shared beyond the session. • Explains informed consent and asks the participant to sign the form while sharing their screen. • Shares the prototype and asks the participant to complete several tasks using the discussion guide. • Monitors messages from observers to follow up on. • Takes notes during the session to remember to explore certain points further.	The researcher marks the participant as attending and sends an egift card to them to say thank you for their time. The moderator sets up a project space in the analysis tool and sets access permissions for the space. They invite collaborators to join. They schedule analysis sessions with collaborators. The testing sessions are spread out over two weeks so the moderator works with collaborators on a rolling synthesis between sessions.	Once the sessions are wrapped, the moderator completes analysis, delivers a report, and presents findings to the team. They also do a data cleanse and move useful data to the research repository. They submit the research assets to the research library, and share the research report across multiple communications channels (sometimes up to 10 channels). They clear sensitive data from their local drive, content production and communication tools, and submit raw data that should be retained to the research repository for safe keeping.
	Observer etiquette form hosted in a survey tool. When the observer signs the observation form, they automatically receive a calendar invite to the research session. Standard-issue calendar system, collaboration, and meeting systems. The moderator is notified via email when an observer signs a form.	Consent form that can be digitally signed. Usability testing to tool to share the prototype, and moderate and record the session. Slack for communication with observers. A whiteboard tool for collaborative note taking during sessions.	Consistently branded egift cards via the usability testing tool or using the egift card platform. Analysis tool must support collaborative analysis across time zones and include access and permissions features. Some researchers prefer a whiteboard tool for analysis.	Research library to which researchers can submit research content. Research data is stored in the research repository for six months before being deleted.
	maze Typeform 31 M	DocuSign maze slack miro	Great Question TREMENDOUS Dovetail miro	KNOWALL MATRIX Dovetail

Depending on the complexity of the workflow you choose, your blueprint could get messy pretty quickly. To give you the space you need to dig into specifics, it's useful to create blueprints that zoom in on more discrete workflows. Say you've created a blueprint for moderated qualitative research, but the flows needed to support informed consent are proving more complex than you thought. To figure out the details, you might create a blueprint that concentrates on describing informed consent only. Whatever level of granularity you choose, to create a blueprint you'll need to:

1. **Choose a workflow to focus on, which will mean interrogating your priorities.** Creating a blueprint is an investment that will pay off in the long term, so you'll want to make sure that you put your energy in the right place.

2. **Do research to understand the phases, experiences, and needs of the people who traverse the workflow,** for example, researchers, PWDR, administrators, vendors, participants, stakeholders, consumers, etc. (For tips on respectful internal research participant recruitment, see Chapter 5.)

3. **Define the various stages of the workflow horizontally at the top of the blueprint.** For instance, if you were mapping the stages of analysis and synthesis, the stages might be "preparation," "migration," "sorting," "analyzing," "synthesizing," "reporting," "sharing," and "hygiene."

4. **Define the tools needed to support each phase and your high-priority needs.** To follow on from the previous example, the tools used for data preparation are likely to be different from those used for synthesizing or sharing research outputs. In an ideal world, the transition from tool to tool should be seamless, but that's not always possible.

5. **Design the front- and back-of-house tooling experience.** How will you let people know where to go, how to access and move between tools, and how to use them well? This work should inform your support systems.

6. **Wherever possible, design in commuter lines for data, too.** You will likely need to work with a data architect to achieve this important goal.

A research tooling blueprint, and especially a collection of blueprints, will help you spot opportunities to consolidate and simplify pathways across multiple tools—or *all* of your tools, which is the ideal. By consolidating finance, onboarding, training, and support operations around your research tooling stacks, you can make operations more efficient (and cost efficient) and simplify the end users' experience, which is always a win.

> **NOTE SEAMLESS SUPPORT**
>
> A key factor in successful tooling is enabling people to use individual tools and move between them to achieve their goals. To achieve this, you'll need to show people that a tool exists (awareness), provide them with access and enough knowledge to get going (onboarding), help them when they first use a tool (first-flight jitters), give them a hand when they get stuck (ongoing support), let them know when something changes (change management), and remove them when they no longer need access (offboarding). For more on delivering scalable support, see Chapter 7.

Data Commutes, Too

There are countless opportunities across the research workflow to make research activities more efficient or effective, or to automate pesky mundanities using data interoperability. But there's a lurking issue. The research technology industry is still maturing, so the kit and kaboodle needed to enable ideal data interoperability isn't yet widely available.

To be fair, perfect interoperability in the technology world is unfortunately rare. When will *all* charge-by-wire devices move to USB-C? Or *all* countries use the same wattage and plug points? As ResearchOps specialist, Casey Gollan, pointed out in his *Medium* post "Open Research Platforms,"[8] "Attempting to work across platforms poses serious challenges. With only limited access to data via manual exports and third-party integrations, it's not practical to maintain ongoing links between research platforms. And without robust APIs, there is no possibility of building something new and innovative on top of your existing research infrastructure."

---

8 Casey Gollan, "Open Research Platforms," *Medium* (blog), June 7, 2022, www.medium.com/@caseyg/open-research-platforms-14b4c7ccf44e

API, SSO, SAAS

Data management comes with a hefty glossary of jargon and acronyms, which you'll pick up over time. Here are three of the most common:

API means *application programming interface*. An API is software code that is used to pass data from one platform to another. In terms of research tooling, APIs are particularly useful in both participant recruitment and respect in research. For example, you could integrate with the company's data lake to enrich a research participant panel with additional data (provided it's consented for that use) or use an API to automate the right to be forgotten (RTBF) in support of data privacy.

SSO stands for *single sign-on* and is usually managed centrally by IT. SSO enables a user to log into one application or network domain with a single set of credentials. IT may require that all the tools you onboard are SSO ready.

SaaS stands for *software-as-a-service*. SaaS is a software licensing model that supports access to software on a subscription basis via external servers—via the cloud, or in high-security settings, a dedicated server. The opposite requires the need to install the software on individual computers.

This is true, but the industry is also maturing rapidly. While robust APIs and standards that support data interoperability aren't yet widely available, vendors are putting in the legwork, and they're often willing to collaborate with research teams and other vendors to meet their needs, provided the result will be universally useful. So, you should define requirements and opportunities for data interoperability, and there are plenty—automated RTBF is a common requirement, for instance—and then work with vendors, data engineering, and data architects to enable as many as you usefully can. In doing so, you'll mature research operations practices and technologies not just for yourself, but for the wider research profession, too.

Foolproof Planning—One Tool at a Time

Big picture thinking (or systems thinking) and interoperability are critical in tooling land, but the robustness of an engine relies on the integrity of its individual parts and selecting and procuring tools is tricky. You might do all the legwork needed to confidently decide that a tool will meet the organization's needs, and then it falls apart in security, finances, or even at contract signing. There always seems to be a long list of unknowns and no way of figuring them out without starting the procurement and approvals journey. It can make you want to tear your hair out, but unknowns (and known unknowns) can be managed by planning, and also with sheer and unabated curiosity and *constant* questioning. The following checklists will help you ask the right questions and do the right things, at the right time:

- Ten key steps to onboard a tool—*any* tool
- Standard tooling criteria
- The power of a question log
- The ResearchOps Planning Matrix

Ten Key Steps to Onboard a Tool—Any Tool

There are entire books dedicated to the topic of onboarding a tool, especially technology, and you'd do well to pick one up. One such guide is *The Right Way to Select Technology.*[9] Although the focus is on technology, their ever-green advice can be applied to tools that are nondigital, too. To onboard a tool, you should do the following:

1. **Craft a business case.** What kind of tool do you need to meet the strategy, and what will it take to deliver it? This includes money, skills, people, and time. A Discovery phase and the ResearchOps Planning Matrix (shared later in this chapter) can help you figure this out.

2. **Get people involved.** Build a selection team or a "research tooling council." A council should include a representative from all your stakeholder groups, like the legal team and IT. The council will help you make better choices than you might have otherwise, and you'll gain the buy-in and support you might need later down the line. You'll also want to get someone in leadership or management on your side.

3. **Decide on selection criteria.** You might build on a set of standard tooling criteria that you develop, or have already developed, for general use. You'll learn more about standardized tooling criteria in this chapter under the unsurprising heading, "Standard Tooling Criteria."

---

9   Tony Byrne and Jarrod Gingras, *The Right Way to Select Technology* (New York: Rosenfeld Media, 2017).

4. **Understand your users' needs.** What are your users' needs specific to the use case? Your user base will often extend beyond just the needs of researchers; you may need to consider partners like privacy, marketing, research participants, or panelists, too.

5. **Gather requirements.** Above and beyond the strategy, standard tooling criteria, and users' needs, what are the technical, procurement, legal, and security requirements specific to this tool?

6. **Ask questions.** There are countless questions involved in delivering a tool, and you should obsessively note them down in a question log, and then obsessively seek to answer them all. The devil is in the details.

7. **Find the right suppliers.** You'll need to find a match between your requirements and the kinds of tools that are available to adapt, adopt, build, or buy. Work within your organization's guidelines for accessing suppliers and think outside the box about whom you consider partnering with.

8. **Try before you buy.** Run a pilot or proof of concept (PoC) with your audience to see how the tools fare on the ground. A test run will do a world of good in terms of speeding and stabilizing decision-making and gaining buy-in both for when you're signing contracts and when you go live. A pilot will also help raise awareness about upcoming changes that will come with the tool. "Awareness" is the first principle of change management.[10]

9. **Negotiate terms.** If the contract is big, let your procurement partners step in to help you partner positively and negotiate the best deal. As noted in Chapter 10, "Money and Metrics," mutually beneficial partnerships should always be the goal.

10. **Make a final selection.** With a final selection made, procurement and onboarding can finally begin. And while procurement is on the go, you'll have plenty to do to make sure that you're ready to roll after contracts are signed and sealed, and to make sure that you've got the resources in place to efficiently support the tool and the people using it in the long-term.

---

10 "The Prosci ADKAR Model," Prosci, www.prosci.com/methodology/adkar

Standard Tooling Criteria

One of the most important assets that you can coopt or create for yourself is a set of standard tooling criteria: a checklist of the must-haves and hard "no's" for tooling compatibility that you can assess tools against without bringing teams of people into the conversation or wasting time on fruitless negotiations. Your organization may already have a set of standard criteria, which you can build on in service of research. If that's not the case, work with partners like IT, procurement, privacy, and security to devise a basic set of standards suited to research. Tooling standards might include requirements to do with accessibility, like Web Content Accessibility Guidelines (WCAG) compliance, security requirements like single sign-on (SSO), access to APIs that support RTBF, certifications to do with data privacy such as ISO27001 or SOC 2, or a requirement from the procurement team that all contracts are signed on your company's "paper."

The Power of a Question Log

Delivering tooling is an endless game of question and answer, which can drive some people to the brink of insanity, but it's a good idea to lean into this trait. In addition to your standard tooling criteria, the more questions you can ask and answer up-front, and en route, the better you can scope the full extent of the work, avoid unpleasant surprises, and make the most of opportunities. While it's unlikely that you'll be able to answer all the questions you have right out of the gate, noting them down will make sure that they're not lost in the tide.

The ResearchOps Planning Matrix

The ResearchOps Planning Matrix (see Figure 8.4 on pages 194–195) is a useful ally to unearth questions that need answering, tasks that need doing, and resources that need securing right up-front (see Chapter 4, "Planning Realistic ResearchOps"). It will help you do whole system planning and account for the requirements of the tool over time. You might store the questions that you generate in a Trello board—you might even use a research insights tool like Dovetail.

Problem Statement or System Name:	Build or Buy What do you need to build or buy to set up operations?
Participant Recruitment	
Knowledge Management	
Onboarding and Support	
Tools and Vendors	
Ethics and Privacy	
Money and Metrics	
Program Management	
People and Skills	
What resources will you need?	

FIGURE 8.4

The ResearchOps Planning Matrix surfaces a complete-as-possible set of needs and questions to consider when delivering a tool.

Standardize

What could you standardize to make operations more efficient or effective? Could anything be automated, documented, turned into a template, or constrained to set criteria?

Specialize and Optimize

Does anything already exist, i.e., tools, partnerships, services, etc., that you could adapt to meet this need? If something already exists, could it be optimized to meet the need even more effectively?

Keep the Lights On (KTLO)

What will you need to do daily, weekly, monthly, and/ or annually to "keep the lights on" and mature the system over time.

The ResearchOps Planning Matrix will help you consider everything that you need to do across the eight elements of ResearchOps to deliver a tool and its adjacent parts.

1. **Participant recruitment** may not be relevant in every situation, but there will often be *something* recruitment related to think about such as:

   • How will participants access the tool and experience it, assuming the tool requires involvement from participants.

2. **Knowledge management** can feel somewhat oblique to tooling projects that have nothing to do with a library or a repository, but there's always a connection. For instance:

   • How will people access the data that's stored in the tool?

   • Will any data need to move into the repository or library and, if so, what is it? How will it get there?

   • What will you need to do to manage the integrity of the data, or archive it?

   • Will you need to build a new taxonomy or specialize the current taxonomy to meet new needs, or are you all set?

3. **Onboarding and Support** can drain operations if not properly planned.

   • How will you make people aware of the tool?

   • What kinds of resources will you need to build for people to onboard the tool and train them in how to use it well?

   • How will you support people in using the tool? Any special support requirements?

   • How will you offboard people when they no longer need access or leave?

   • You'll need to consider change management, too. How will you bring people on the journey and create a sense of community and proficiency?

4. **Tools and vendors** are at the heart of the work you're doing, but there are standard technology-related questions that you should ask:

   • How will you migrate data in and out of the tool?

   • Does the tool fit within your current research tooling stack and workflow?

   • Is the tool mature enough to deliver the strategic vision, or will you need to influence the vendor's roadmap to achieve your goals?

   • How will you manage the vendor partnership, and do you get a good vibe so far? Good tooling is all about partnering.

   • Do you require other partnerships for the tool to be successful, say engineering, and does a partnership look promising?

5. **Ethics and privacy** is important to consider every time that you onboard a tool. The following ethics and privacy questions should be included in your standard tooling criteria.

   - Will the tool pass muster in terms of data privacy and security? You'll need to get approval for its use, so make sure to involve privacy and security early on. You don't want to waste other people's time considering a tool until you've got at least a broad approval from these important partners.

   - Will compliance needs impact the researchers' or participants' experience so that the tool is no longer workable? A well-known diary study tool retains ownership of participants' data, for instance, which isn't a good participant experience.

   - Is the tool a good choice ethically? Are there any concerns about the organization's working practices? Certain transcriptions tools, for instance, have had a bad rap in the past for poor employee practices.

   - If necessary, is the tool accessible to people living with a disability? This should be a standard criteria for *all* tooling. Unfortunately, many research tools aren't yet accessible.

   - Who will you need to work with to get approval and what do you need to do to get on their roadmap? Privacy teams tend to be busy, so make sure that they have availability to work with you.

6. **Money and metrics** are another always-on requirement for delivering tools. You'll need to ask:

   - How will you secure the funding needed to onboard and maintain the tool?

   - How will you fund any adjacent requirements of the tool? For instance, pay-as-you-go tools demand ongoing financial management, so you'll need bandwidth to manage that administration in the long run.

   - Is the vendor contract financially viable for everybody? If you intend for the vendor relationship to last, you'll want to make sure that the vendor values the relationship financially. I've been known to bargain startups *up* to meet that criterion.

   - How will you measure the return on investment (ROI) or improvements in efficiencies that the tool will provide? In the case of a participant recruitment platform, perhaps you'll measure shifts in the turnaround time for completing a recruit.

   - Finally, is the tool financially scalable? In other words, is the per-seat cost prohibitive to scale?

continues

7. **Program management** is often forgotten in planning. Take time to consider:

   - Are there any restrictions that might require forward planning regarding who uses the tool and how it's used? For instance, in the case of an unmoderated research tool, is there a restriction around the number of studies that can be run concurrently at any one time?

   - Will you need a heads-up about how a tool is being used, or how much it's being used, to make sure that it's set up to meet a demand?

8. **People and skills** topics are often left last in planning efforts, if considered at all, but helping people use your tools, and use them well, is a critically important effort. Good questions to ask:

   - Does the tool fit the target audience's research capabilities and skills maturity? If not, will you need to offer specialized training or extra support?

   - How will you manage permissions, and who has access to the tool?

   - Will you need to produce and maintain guides, guardrails, or templates to enable the correct use of the tool? If so, what kinds of templates? And are researchers available to help you develop and test them ahead of the rollout? Could the vendor help with this?

   - From a strategic point of view, will the tool encourage research behaviors and practices that will positively impact research quality? Who can help you monitor that, and how will you measure it?

   - Will the tool impact the day-to-day work lives of researchers? For instance, will the tool create a demand for senior researchers to approve activities or to support people to use it?

The Good, the Bad, and the Ugly of AI

Since the launch of OpenAI's ChatGPT in November 2022, conversation about the good, the bad, and the ugly of artificial intelligence has taken over conversation—everyone from homemakers to technologists has an opinion about what AI means to the future of humanity. The promise of AI's capabilities to augment, even if not outright replace, human intelligence is undeniable. Considering that research is a knowledge service, it would be irresponsible not to keep tabs on AI developments and leverage this ultimate form of automation within your operating systems. But, as ChatGPT agrees, it isn't yet capable of replacing researchers (see Figure 8.5).

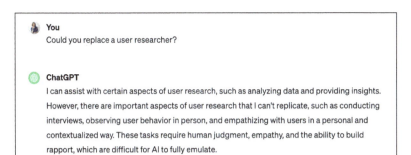

You
Could you replace a user researcher?

ChatGPT
I can assist with certain aspects of user research, such as analyzing data and providing insights. However, there are important aspects of user research that I can't replicate, such as conducting interviews, observing user behavior in person, and empathizing with users in a personal and contextualized way. These tasks require human judgment, empathy, and the ability to build rapport, which are difficult for AI to fully emulate.

FIGURE 8.5
Researchers might question ChatGPT's assertion that it can analyze data and provide insights, and I agree. A human should *always* be in this loop.

The company User Interviews (UI) often hits the proverbial nail on the head when it comes to content. In 2023, they published a report called "We Surveyed 1093 Researchers About How They Use AI—Here's What We Learned,"[11] and it's entertaining and informative. The report states "The researchers in our audience listed increased efficiency as the top benefit of AI. But does this efficiency come at too great a cost?" The answer in part: 'Here's a handful of the words and phrases that participants used to describe AI's outputs: "hallucinations," "wonky," "absurd," "nonsense," "embarrassing," "really poor (and frankly, hilarious)," "like a junior assistant," "like a seventh grader," "like a glorified Google search," and "basically useless."' So, although AI is not yet capable of replacing researchers, it can increase efficiency.

How and when to use AI is dependent on the individual tools in your tool stack, organizational priorities, and what you want to achieve. Common use cases include automating notetaking during user interviews, generating transcriptions and translations, or eliminating the blank page when writing reports. Treat AI as you would any other tool: question where it will be best used, and how it will impact the culture, results, and value of research. As Chief Scientist for IBM Grady Booch[12] said, "A fool with a tool is still a fool."

---

11 "We Surveyed 1093 Researchers About How They Use AI—Here's What We Learned," User Interviews, August 2023, www.userinterviews.com/ai-in-ux-research-report

12 Grady Booch is IBM's Chief Scientist for Software Engineering and for Watson/M at IBM Research, so he knows a thing or two about AI. Watson, an early pioneer in AI, remains a significant player in the field today.

The ResearchOps Planning Matrix will also help you define questions across the three phases of operational maturity.

1. **Build/buy and standardize:**

   - Will you build or buy the tool?
   - *Why* are you delivering a tool in the first place? Every tool should have a well-defined purpose and predefined place within the research operating system. It should help to deliver value.
   - Or does a tool already exist within the organization, which you can adopt or optimize to suit the context?
   - Apart from the tool itself, what services and resources will you need to buy/build and what could you standardize to support the tool and the people who use it?
   - How will you gather feedback, measure usage, and track success?
   - What kind of capabilities should the tool offer to enable you to track metrics?

Today, there are an abundance of purpose-built research tools from which to choose, so building your own is often of limited value. If do want to build a tool in-house, note the size of the companies who offer the service commercially. If you think you can build and maintain the service internally with fewer people and less effort, you're either in the wrong business, or you're ripe for a rude awakening! Tools that are built in-house, like their commercial counterparts, require significant ongoing investment to deliver. If you choose to take this route, assume that evolving and maintaining the tool will become your full-time occupation—unless you hand the responsibility over to someone else who shares your intentions. In short, if a commercial service is available, buying a tool is generally more cost-effective than building a tool: it's lower risk, and offers an easier exit.

2. **Specialize and optimize:**

   - Are there any capabilities you'll need to optimize and specialize the tool for your context? For instance, can you embed specific help tips or callouts, set granular permissions levels, add custom-built templates, brand the tool, or integrate with APIs?

   - After the tool is onboarded and rolled out, what does success look like? The answer to this question will inform what you track.

   - Are there specific metrics that you want to move the needle on?

   You may only be able to answer these questions when the tool and attendant services have settled down, but by giving them thought during planning, you'll strengthen your chances of building/buying and standardizing a tool that will grow with you in the long-term.

3. **Keep the lights on:**

   - What are the daily, weekly, monthly, and annual tasks that you'll need to do to keep the lights on?

   - What will you need to do to provide logistical and craft support, if you choose, in the long-term?

   - Do you plan to use the tool for one contract year, or many years into the future?

   - Will the tool scale in relation to your growth forecast: perhaps it suits 50 researchers now, but should it be ready to scale to 100 researchers within a year or two?

   - What does the financial forecast look like to maintain the tool, and is it viable?

   Onboarding and offboarding a tool takes investment, so the predominant aim should be to set yourself up for a multiyear relationship.

In a Nutshell

However well you plan, your research tool stack will likely always be in flux. At points, you may need to swap tools out: a vendor may take their tool in an unexpected direction, perhaps there are problems with quality or reliability, the price per seat goes up unexpectedly, or a tool is abandoned because researchers go collectively sour on it. I've seen all of these happen. It's at these moments that your well-documented tooling strategy, maps and models, question log, standard tooling criteria, and strong partnerships with the procurement, security, and IT teams will become invaluable. However simple or innovative the tool, these are the things to remember:

- **Before doing anything, consider the strategic intent of the tool, how it will increase the value that research provides, and then users' needs.** This will inform the type of tool that you procure, and how you should embed it into existing technologies, systems, and cultures.

- **Create a research tooling blueprint.** You might map the tool stack for your entire research operating system or for discrete workflows, like the end-to-end management of informed consent, submitting reports to a library, or doing a remote unmoderated research study. This will help you understand technical or informational gaps, pain points, support requirements, and opportunities for automation or interoperability.

- **Develop reusable assets like standard tooling criteria and protocols for tracking a question log over time.** Also, use the ResearchOps Planning Matrix to consider the eight elements of ResearchOps and the ongoing requirements of onboarding a tool up-front. A tool (and everything it takes to provide it) is rarely a small investment, so planning should be on point.

Respect in Research

If you've ever been on the receiving end of poor privacy practice, you'll know how invasive it can feel. As mentioned in an earlier chapter, I was once asked to chat to colleagues from another team about my experience using their product, only to discover the following day that the "chat" had been recorded without my knowledge and published on the company wiki as a "research interview," which misconstrued the context of the conversation entirely. If I'd said something untoward, it would have put me in a very vulnerable position. It's rare for a bona fide researcher to make this kind of rookie error, but, intentional or not and whatever the scenario, research opens people up to being harmed.

Researchers collect snippets of people's lives—their memories, dreams, needs, frustrations, and joys—and record it as data. Every interaction or experience shared during research is mediated through data, from participant recruitment, to consent and recording, analysis and synthesis to the final knowledge store where, without data, what was once substantiated knowledge may quickly become folklore. But data isn't soulless information stored in digital or analog files. It has something of a radioactive charge, and it lives well beyond a research study's final hurrah, potent in its potential to deliver both help and harm. Managing how data is acquired, used, and maintained throughout the research workflow, and how people experience taking part, defines the specialist work of respect in research: research that's kind, honest, responsible, and legally compliant.

Respect in research is important and evolving work, and it will impact every part of your research operations and the research experience. To do this work well, you'll need to be aware of emerging ethics and privacy themes, particularly as technology evolves and the research profession gains velocity—and as you scale research, which means scaling your risks and responsibilities. You'll need a familiarity with key ethics, legal, and data governance principles and terminologies, and, crucially, you'll need an open-minded curiosity. Surprisingly, respect in research is less about checklists and legalese, and more about codesign, debate, and practical philosophy.

The Respect-in-Research Trifecta

There are three aspects for delivering respectful research, each as important as the other, and each equally reliant on the other (see Figure 9.1):

- **Ethics:** How are people (participants, researchers, and others) respected and cared for before, during, and after they take part in research?

- **Data governance:** How are different kinds of personal data collected, used, accessed, maintained, and safeguarded throughout the research workflow?

- **Data privacy:** Are research practices compliant with privacy law, and do they pose a legal or reputational risk to those involved?

FIGURE 9.1
The respect-in-research trifecta encapsulates three equally important parts: ethics, data governance, and data privacy.

To successfully deliver respectful research, whatever your scale, you'll need to address and orchestrate all parts of the trifecta synchronously. This doesn't mean, though, that you'll need to become (or hire) an ethicist, lawyer, or data specialist to succeed. Instead, as with most things in ResearchOps, your primary work is in gathering and coordinating existing expertise. In this case, your organization's legal, security, and IT teams will all have high-stakes interests in making sure that research practices meet their needs, including laws and regulations. As ResearchOps, you'll also need to make sure that while researchers meet compliance expectations, they are also empowered to do their best research ethically. With thoughtful design, compliance needn't be a blocker.

Informed Consent—More Than Just a Form

Mention the words *research ethics* or *data privacy,* and *informed consent* is often the first thing that a researcher will think about. While informed consent is not the be-all and end-all of respectful research, it's worth mentioning up front because it sits at the heart of the respect-in-research trifecta (see Figure 9.2): between what is right to do (ethics), how to handle data so that the agreement is honored (governance), and what you're required to do (regulation and law).

FIGURE 9.2
Informed consent sits at the heart of the respect-in-research trifecta. It communicates how you will treat a participant and their data, and the operations that you must have in place to stand by the agreement.

Informed consent should be more than just a form, it should be a *practice* that's integrated into research workflows—and researchers' mindsets. To deliver an informed consent practice that's legally compliant and ethical, you should:

- Work with legal partners to create consent forms. A consent form is a legally binding agreement between the participant and the organization, and it must be written correctly.

- Pull out your best collaboration skills to make sure that agreements are written in plain language so that people from all walks of life can understand what they're signing up for. Consent must be *informed*, after all.

- Keep the form as short as possible so that it's not overwhelming. A two-page consent form is entirely possible, and there's plenty of precedence. Even a one-page consent is possible, depending on the context and the needs of the research.

- The form should be specific to the type of research and situation. It's not unusual for a mature research team to have multiple types of consent forms for various scenarios, like research with employees or accompanied minors. You may also provide templates that can be customized without breaking the legal integrity of the contract.

- Make sure that your consent form enables participants to explicitly select options for how they interact with you. For instance, when they sign the form, allow them to select (or consent) to being video recorded, audio recorded, or whether they consent to people observing the session. (Note: When a participant consents to being observed, let them know how many observers you expect to attend. Inviting a crowd of people to observe without warning can be an unnerving experience for the participant— it's not kind.)

- Your consent form should be accessible to people living with disabilities. If the form will be used to gain consent during accessibility research, it should include clauses that cover accessibility research specifically, which requires special provisions.

- If research regularly involves participants whose first language isn't your organization's home language, provide the form in their language.

- If you research in other countries, work with legal partners to make sure that the agreement is compliant with those countries' privacy regulations. Rules change from country to country.

- Send scheduled participants a preconsent in their native language: an information sheet that includes a PDF version of the consent form so they can read it ahead of time and ask questions or cancel their participation, if necessary. It's often required that a consent form be signed at the time of the session and not before but check in with your legal counsel.

- Provide researchers with a script to use as a verbal consent at the start of each session. The script should include a verbal cue that lets the participant know when recording has commenced.

- You don't want to make a participant unnecessarily nervous by laboring over their data rights, but it's good practice to briefly remind people about their consent at the end of the session, and how to make contact should they want to revoke their consent post-session. You could wrap this final note up with a courteous message, sent with the participant's thank-you gift.

You can read more about integrating informed consent into the participant recruitment workflow in Chapter 5, "People to Take Part in Research."

NOTE **NDA VERSUS CONSENT**

A nondisclosure agreement (NDA), which is also sometimes called a confidentiality agreement, is often included in a participant consent form. An NDA asks participants to contractually agree not to share information, including any intellectual property (IP), which they may learn about as a result of taking part in research. Some people confuse an NDA with consent, but they're not the same. While an NDA can be included as part of a consent form, and often is, it's not a substitute for a participant consent form. This is because an NDA protects the company and *not* the participant.

Research Ethics: Setting the Record Straight

There are several important themes that are often missed or misunderstood about respect in research and understanding them will help you adjust how you approach this work for the better. Sometimes researchers, even senior research leaders, may want to curtail your ethics efforts because, "A consent form is all we need." And legal partners may fail to understand why you're pressing them to make plain language of hard-to-read legalese. Even people working in ResearchOps may fall short simply for lack of exposure to these themes. These are the three key themes that you must know about:

- *Legal* doesn't mean *ethical*.
- Ethics is largely unregulated.
- Ethics is needed in *all* research settings.

Legal Doesn't Mean Ethical

In January 2012, Facebook data scientists and researchers ran a one-week experiment to explore whether "emotional states can be transferred to others via emotional contagion."[1] As part of the study, Facebook skewed the newsfeeds of almost 700,000 of its users so that some users saw content that was happier, while others saw content that was sadder. When the week-long experiment was over, it was discovered that people who'd been shown happier content posted more positively, while those who were shown negative content posted more negatively.

At face value, the study was an A/B test, a common research method you're likely familiar with. A/B tests are regularly run by companies of all sizes all over the world, and often without explicit *or* informed consent from their users. Instead, consent is obtained via the organization's terms of service or privacy policy, which users agree to when they sign up to the service—you've likely signed it yourself. In this case, Facebook carried out the research under its 9,000-word terms of service agreement, which, from a purely legal perspective, was within compliance. From an ethical standpoint, however, the study was described as "creepy," and the story hit the headlines.

Professor Susan Fiske, editor of the study and professor of psychology at Princeton University, said in an interview with *The Atlantic:*[2] "It's ethically okay from the regulation's perspective, but ethics are kind of social decisions. There's not an absolute answer. And so, the level of outrage that appears to be happening suggests that maybe it

---

1 Adam Kramer, Jamie Guillory, and Jeffrey Hancock, "Experimental Evidence of Massive-Scale Emotional Contagion Through Social Networks," *PNAS* 111, no. 24 (2014): 8788–8790, https://doi.org/10.1073/pnas.1320040111

2 Adrienne LaFrance, "Even the Editor of Facebook's Mood Study Thought It Was Creepy," *The Atlantic*, June 29, 2014, www.theatlantic.com/technology/archive/2014/06/even-the-editor-of-facebooks-mood-study-thought-it-was-creepy/373649/

shouldn't have been done...I'm still thinking about it, and I'm a little creeped out, too."

Although just over a decade old, Facebook's story is still relevant, and it shares lessons from which ResearchOps teams can learn. Chiefly, legal compliance isn't the hallmark of either low-risk or ethical research. *Ethics needs more.*

Ethics Is Largely Unregulated

While all organizations are bound by increasingly more stringent privacy regulations, like the General Data Protection Regulation (GDPR),[3] California Consumer Privacy Act (CCPA),[4] or Brazil's General Data Protection Law (LGPD, in Portuguese), private companies aren't required to agree to the same ethical research standards that government and academic institutions do. While government or academic-funded research in the U.S. must be approved by an Institutional Review Board (IRB), guided by a code of conduct called the *Common Rule*, research done within commercial contexts has no ethics rules. It must self-regulate. Though different countries and contexts have different regulations for scientific or academic human research, commercial user research isn't formally regulated anywhere on earth.

Many science-oriented, for-profit industries, like drug or wellness companies, voluntarily use the Common Rule, or something similar, to make sure they have ethical due diligence in place—often for reasons of risk and brand management as much as anything else, if one is to take a cynical view. But most commercial organizations who aren't scientifically or socially oriented have no ethical framework or have something minimal: a code of conduct, perhaps, that's loosely adhered to. If this sounds like you, you're not alone. Ethics are commonly underfunded in user research due to the following reasons:

- There's a general lack of understanding about research ethics within the broader user research profession. It's often assumed that legal compliance, a one-page code of conduct, and a consent form constitute an ethical practice. They don't.

---

3   "Complete Guide to GDPR Compliance," GDPR.EU, www.gdpr.eu

4   "California Consumer Privacy Act (CCPA)," Office of the Attorney General, www.oag.ca.gov/privacy/ccpa

GDPR, LGPD, CCPA, AND THE COMMON RULE

GDPR stands for the *General Data Protection Regulation,* a ground-breaking privacy regulation that was introduced in 2018 by the European Union (EU) to protect the personal data of its residents. If your organization handles EU residents' data, you must comply with the GDPR. Critically, the GDPR puts individuals in charge of their personal data. The GDPR is the most extensive privacy law in the world and has set the bar high globally for privacy regulation. Many countries have updated, or are working toward updating, their privacy laws to comply with the GDPR's strict requirements. Most notable of these is Brazil's General Data Protection Law: Lei Geral de Proteção de Dados Pessoais[5] in Portuguese or LGPD, for short.

The **LGPD** came into effect in September 2020, and it largely mirrors the GDPR. It aims to unify 40 different Brazilian laws that regulate the processing of personal data and protect the collection, use, processing, storage, and transfer of the personal data of Brazilian residents, as the GPDR does for European residents and the CPRA does for California residents.

CCPA stands for the *California Consumer Privacy Act,* which is the State of California's privacy law. It has been in effect since 2020 and was amended to include additional privacy protections (called the *CPRA* or *California Privacy Rights Act*), which came into effect on January 1, 2023. It's worth mentioning the CCPA because, although its protections apply only to Californian residents, it's the strictest privacy law in the U.S.

To explore the laws or regulations of countries around the world, CNIL's[6] interactive map is a useful resource: www.cnil.fr/en/data-protection-around-the-world.

The Common Rule[7] is the common term for the wordier Federal (U.S.) Policy for the Protection of Human Subjects, 45 CFR part 46. It's a rule of ethics that applies to biomedical and behavioral research involving people, and it's the baseline standard of ethics that all government-funded research must follow. Most academic institutions in the U.S. hold their researchers to the Common Rule, regardless of the source of funding.

---

5   "LGPD English Version," lgpd.brasil.com.br, June 7, 2019, https://www.lgpdbrasil.com.br/?s=english

6   "Data Protection Around the World," The "Commission Nationale de l'Informatique et des Libertés" (CNIL), www.cnil.fr/en/data-protection-around-the-world

7   "Federal Policy for the Protection of Human Subjects ('Common Rule')," U.S. Department of Health and Human Services, www.hhs.gov/ohrp/regulations-and-policy/regulations/common-rule/index.html

- Because commercial research contexts aren't required by law to follow an ethics process and there are no fines, ethics efforts are often deprioritized (if they're considered at all).

- The investment needed to support ethical research practices is often scant. Even with the best of intentions, there's always something seemingly more important to put time and money toward.

- Ethical situations rarely emerge, so it's easy to turn a blind eye. You may only be able to prove the value of investing in ethics after the organization suffers reputational damage of some kind. If this does happen, leverage it for extraordinary progress. This is your moment to shine a light on the need for better ethics practices!

It's a list that paints a pretty dismal view, but there's a lot that you can do to counter these issues, and you won't need an army of specialists. (See "How Ethical Is Your Research?")

Ethics Is Needed in *All* Research Settings

It's often assumed that ethical practices are only relevant in contexts where obviously sensitive topics are being researched, such as illness, mental health, bereavement, criminal history, poverty, and abuse. Or when research involves vulnerable people such as children, people with cognitive or physical impairments, or people with low literacy. It's true that these topics and contexts make an outright demand for sensitively executed research, but *every* organization, whatever the context, must take ethics into account.

Here's a scenario: A participant takes part in a usability study about software they use as part of their job. They relax during the session and openly share information about a confidential project, forgetting that they've consented to being recorded. Once the study is complete, personally identifiable data from the session is shared across the company to build empathy for the use case, the details of which get back to the participant's boss, and they're called out for breaking confidence. Their job's on the line. They signed the consent, so the legal aspect is covered, and the researcher did everything right, but the participant's still been harmed by the experience, and the researcher feels terrible. Of course, it wasn't their fault. The work of operations isn't only to prevent harmful scenarios from happening—even with best efforts, things can go wrong—but to have protocols in place for remedying issues when they do arise.

RANDOM ASSOCIATIONS AND CHANCE

At a conference years ago, a speaker shared a story about research that they'd done involving traffic signs, if I recall correctly. During a session, the researcher showed a participant an image of an unremarkable intersection, and they broke down. By complete chance, they had been involved in a road accident that had been fatal for a pedestrian at that very intersection. It's an extreme example, but it illustrates the point.

Even if the research subject seems vanilla to you, the experience can be triggering. For example, if the research topic is social consumption of alcohol, the research could involve participants who have a history of alcoholism. Dieting or exercise could trigger issues around body image. Mindfulness may involve people struggling with anxiety. The national lottery may involve people who are struggling financially. Participants tend not to know what they'll cover in a session and, therefore, how they'll react. In short, don't make assumptions and design ethics practices into every research system.

In a talk about research ethics for the Cha Cha Club,[8] IDEO's Design Research Operations Director, Leah Kandel, shared that they regularly hire subject-matter experts like historians to help the team understand the historical context of the topic that they're researching. This helps the team to approach the research, and the participants, not only more productively, but with more sensitivity. It's an impressive approach.

How Ethical Is Your Research?

It's not uncommon to hear ResearchOps professionals lament: "I've written an extensive ethics or data-handling playbook, but people are ignoring the guidelines." (To be fair, this is a common complaint about most written material!) It's not that people are innately uncaring. Instead, in the speedy humdrum of getting things done, and even with the best of intentions, making decisions about what is right versus wrong can be tricky and time consuming. The *Little*

---

8   The Cha Cha Club is a members' club for full-time ResearchOps professionals, which I founded and run to this day. You can watch Leah's talk here (it starts at 2:35): www.youtube.com/watch?v=4PGRIq14erY&t=9s

First, seek to understand your ethics context. Ask questions like:

- Does your context mean that you have explicit ethics requirements that you must adhere to, such as the Common Rule or another code of conduct?

- How often do researchers engage with obviously sensitive topics or vulnerable populations? Some topics might not seem obviously sensitive, so think carefully about the things researchers might ask participants to experience.

- Have researchers had concerning ethical experiences in the past, and what went well or not-so-well in resolving them?

- Do researchers have particular ideas, needs, and worries about ethics in research?

- Do your legal partners have any advice, thoughts, or resources regarding ethics?

- If your organization has a market research team, or any other team involved in doing research, do they follow a particular code of conduct?

Book of Design Research Ethics (the first edition) by IDEO[9]—they're an inspiration when it comes to ethics—puts this beautifully: "Respect, responsibility, and honesty sound great. But they're big abstract ideas that seem completely clear until we're asked to define and apply them in the complicated, messy, human situations of real life." (In 2024, IDEO released a second edition of this enduring resource.[10])

So how do you turn the big abstract ideas of ethics into day-to-day habits? Even if you're a team of one, or a researcher who wants to make strides on the side, there's a lot you can do to shift the ethics needle, and the best place to start is by doing an ethics audit. An ethics audit will help you understand how your state of play compares with best practices, and the systems you'll need to deliver to get ethics on track. See the "ResearchOps Ethics Checklist," which follows.

---

9 *The Little Book of Design Research Ethics*, IDEO, 2015, www.ideo.com/post/the-little-book-of-design-research-ethic

10 *The Little Book of Design Research Ethics*, 2nd ed., " IDEO, 2024, https://page.ideo.com/lbodre

A RESEARCHOPS ETHICS CHECKLIST: AUDIT YOUR PRACTICES

Next, audit your practices (these are baseline requirements, whatever your context):

- Do you provide support, written guidance, and training so that researchers understand how to conduct research ethically and what to do when difficult circumstances arise? It's also essential to make research observers and consumers aware of research ethics, and how to treat people and their identities.

- Do you provide an ethical research checklist to help researchers deliver respectful research in each and every study? You can find examples of ethics checklists for researchers on page 219. (A researcher's checklist is different from this ops-focused checklist.)

- Do you have an appropriate ethics code of conduct like the U.S. government's Common Rule or the UK's Market Research Society (MRS) code of conduct for researchers to use? If you're not required by law to adopt a regulated code, you could work with researchers to make your own—they're often wordy and outdated.[11]

- Do you have processes in place—light or heavier, as appropriate—to help researchers vet whether the potential benefits of a study justify participants' time and levels of vulnerability? You might build this assessment into your research prioritization framework as detailed in Chapter 11, "Getting Priorities Straight."

- Do you give researchers access to connections and resources that support ethical research when they have questions, even before doing research, like access to a legal consultant, ethics expert, or a review board?

- Do researchers habitually offer participants a clear explanation of why you are doing the study and what's required for them to take part? Researchers may sometimes need to omit or obscure details about a study, but deception and lying should be avoided at all times.

continues

---

11 The UK's Market Research Society (MRS) released a new code of conduct in May 2023, and it's to the point (23 pages) and easy to read. www.mrs.org.uk/pdf/MRS-code-of-conduct-2023.pdf

- Are participants aware that they're taking part in a research study? Are they informed of how their personal information will be used? In other words, do you have an appropriate informed consent practice in place that researchers and participants are comfortable using—*and* do use?

- Do participants understand that they can ask at any time that you delete the PII you hold about them as a result of taking part in research? (The legalities of consent withdrawal versus RTBF [Right to Be Forgotten] can get tricky[12] and depend on the contents of your consent form and your jurisdiction, but you should work with your legal partners to define protocols that ensure participants' peace of mind and rights at all times.)

- Do researchers and observers conduct themselves in ways that are respectful of participants from all walks of life? How do you help them be aware? And what happens if they aren't? You'd be surprised at how often this ethical standard is broken, even by people who are ordinarily "nice."

- Do participants leave a research study the same or better for having taken part? You could monitor this via a post-study survey, and report and monitor metrics via your quarterly operating report as detailed in Chapter 10, "Money and Metrics."

- Do you have reliable and accessible resources to support participants (and researchers and observers) should something upset or disturb them during or after the research session or study? Perhaps a counseling service or support in accessing helplines?

- Is there a need to maintain the anonymity of participants above and beyond the requirements of law in all or some circumstances? Do you have guidance and processes in place to enable this? For instance, do you provide software that enables video blurring and accesses permissions for files?

- Do you have an easy-to-use process in place that enables participants to communicate a change in consent or the right to

---

12 Ben Wolford, "Everything You Need to Know About the 'Right to Be Forgotten,'" GDPR.EU, https://gdpr.eu/right-to-be-forgotten/

be forgotten (RTBF), even well after they took part in a study? In privacy-progressive countries, RTBF is often a regulatory requirement.

- Do you have data management processes in place so that you can locate and delete a participant's personal data—*all* of it—should they ask? This sounds easy, but the technical aspects can be tricky. Again, in some countries, honoring the RTBF is required.

- Do you provide field support for researchers who are visiting locations on their own? For reasons of safety, it's not unusual to make sure that researchers always travel in pairs and have access to communications technology at all times and contact information for how to get help.

- Do you provide regular debriefings or prearranged counseling sessions for researchers who are engaging in sensitive and challenging topics?

- Are researchers encouraged to plan in "firebreaks"—gaps between potentially stressful research studies—to avoid emotional burnout?

- Do your participant recruitment practices enable and support diversity and inclusion? The UK's Equality Act 2010 describes diversity and inclusion as "age, disability, gender reassignment, marriage and civil partnership, pregnancy and maternity, race, religion or belief, sex, sexual orientation."[13] These are often seen as sensitive categories, so it's especially important that you work with your legal (and ethics) counsel to define how you handle these categories of data. Definitions for PII and Sensitive Personal Information are shared later in this chapter.

- Additionally, do your research practices and infrastructure support income diversity? For instance, can research participants engage with iOS and Android devices, which tend to be more affordable? You may also support travel costs for attending in-person research.

---

13 Equality Act 2010, The National Archives, www.legislation.gov.uk/ukpga/2010/15/section/4

Small but Powerful Efforts

The big and important requirements listed in the ResearchOps Ethics Checklist are...*big and important.* But there are countless and equally powerful opportunities to infuse respectful practices throughout the research workflow and, critically, within research culture and awareness. For example, you could:

- Look for opportunities to build one-sentence tips into tools and templates that remind researchers to double-check their actions when they're about to share participants' personal information.

- Inaugurate an ethics council: a group of committed researchers and other stakeholders, like lawyers, who work together to help make decisions about ethics and, as vocal champions, help make ethical research a day-to-day habit.

These tips have been mentioned before, but they're worth repeating:

- Provide researchers with a template message to share with participants at least 48 hours prior to a session—you might even automate it. Apart from providing session details, give participants the opportunity to review the consent agreement in their own time alleviating the pressure to sign unsighted because people are waiting and watching.

- Include a standardized comment in a post-session thank-you note, which might include an egift card, to remind participants that they can make contact (and how) if a concern or question should arise. You might also invite participants to complete a survey (keep it short and optionally anonymous) to let you know about their experience. Even small discomforts are worth ironing out.

It's useful to map these kinds of tactics against your research workflows. Prioritize the best of them to include in your operational strategies—if they're low lift, just get them done—then get to work and make your ethics priorities a reality.

> **NOTE** THE COMPLIANCE ALLIANCE
>
> There are excellent examples of tight partnerships with legal and security professionals within the ResearchOps profession. One such partnership was even branded "The Compliance Alliance," a standing partnership between research operations, researchers, and legal partners that ensured an ongoing collaboration. As with most things in ResearchOps, respect in research is rarely "done," so your relationship will be ongoing.

Be Responsive, Not Mechanistic

While repeatable or automated mechanisms are always useful operationally, it's worth remembering that at the heart of ethics lie thoughtfulness and sensitivity. Empathy is lived and dynamic, so make sure that your ethics practices are equally responsive. For instance, at the outset of COVID-19, ResearchOps professionals across the world adjusted participant recruitment messaging to acknowledge the stressful circumstances that many participants found themselves in. For instance, Noel Lamb, who was Senior Manager Research Operations at Salesforce at the time, crafted this recruitment message:

> Feedback from our participants has always been at the heart of everything we do. We understand the challenges some are encountering as COVID-19 surfaces around the world and realize this may not be the right time for you to participate in one of our research studies. If you'd like to be removed from our invite list, even temporarily, please let us know. The health and safety

of our customers will always be our most important consideration, and we commit to supporting your needs to the best of our ability.[14]

Or perhaps you're doing research with small businesses that are battling events out of their control—farmers during a drought or small businesses in the midst of a financial crisis. The need to be sensitive, prepared, and responsive abounds, so revisit your operations regularly to make sure that they're context appropriate.

Data Everywhere: Governance Is Key

I used to run a half-day workshop that had attendees map out the journeys of the research data that they collected—the *real* journey—and it was always messy. Even in organizations with relatively stringent data security protocols, stick drives or memory cards ended up lost under sofas, files were stored on local hard drives, and identifying transcripts gathered in desk drawers. Even in highly digitized environments, data has a habit of dispersing.

Research data, which is distinct from the anonymized notes or research reports generated during research and curated as part of knowledge management (see Chapter 6, "Long Live Research Knowledge), includes information of any kind that's gathered throughout the research workflow. In the digital realm, this might include survey responses, demographic data used in participant recruitment, digital consent forms, audio-visual recordings, images, and transcriptions. In the physical realm, data might include photographs, addresses, and paper consent forms. Whatever data you collect, to make sure it is properly handled, you'll need to define and manage the following:

- How data is created or collected.
- Why it's collected is equally important: The purpose or use of the data, governance around how it can be used, and how long it should be retained to meet that use.

---

14 Kate Towsey, "What Research Ops Professionals Are Doing in Response to COVID-19," *Medium* (blog), April 6, 2020, www.medium.com/researchopslife/things-research-ops-professionals-are-doing-in-response-to-covid-19-ee0f-0b5e3ea6

- Where and how the data is stored: *data at rest* in information security parlance. Usually, this kind of data is stored in a research repository as shared in Chapter 6.
- How data is transported from one location to another: *data in transit.*
- Who can access the data: *permissions.*
- How the data is structured: *data architecture.*
- The terminology you'll use to define the data and make it findable: *taxonomy.*
- How you'll manage the quality and integrity of the data so that it's trustworthy and useful—for instance, in quantitative research or if you're managing an in-house participant recruitment panel.
- How you'll archive data that's no longer in use or delete data that shouldn't be retained, thereby reducing risk.
- How you'll enable RTBF or *data subject access requests* (DSAR). DSAR describes when someone requests to access or understand the scope of the personal data that an organization has collected, used, and stored about them.

Data governance is an evolving specialty that involves multiple expertise, including lawyers, information or data security specialists, data analysts and engineers, and IT. If there isn't a discrete data governance team with your organization, data governance tasks will usually be either loosely or cohesively covered cross-functionally.

Mature research organizations rely on excellent data management, not just from a compliance point of view, but to build and maintain infrastructures and interoperability for best practices in both participant recruitment and knowledge management. So, the need for data governance skills goes well beyond the needs of respect in research. While you can deliver data governance by building rock-solid partnerships with specialists in other teams, ideally, you'll hire a data governance expert to own this valuable responsibility, which should include managing a research repository.

LEARN THE LANGUAGE OF DATA GOVERNANCE

2FA stands for *2-Factor Authentication*. **MFA** is the alternative and stands for *Multifactor Authentication*. It's likely that you already use 2FA to log in to online financial services or consumer services like Amazon. 2FA increases the security and traceability of who is logging into a service. The first factor or authentication is your password, and the second factor may be a text including a code sent to your mobile phone, biometrics like a fingerprint or retina, or a code using an authentication app. If you're procuring a tool that researchers will need to access, you may be asked to require 2FA as part of standard requirements.

Data at rest describes data that's inactive. In other words, it's not being used, and it's not in transit. Data at rest might include data stored on a database or drive, whether on a server or a hard drive. This is the least vulnerable data state, as long as the database in which it's housed is secure. Your information security specialist may ask your vendor to provide a technical diagram to illustrate where data is in transit and where it is at rest across a platform.

Data in transit describes data that's moving from one location to another whether via email, a messaging service, collaboration tool, or a network—even via a stick drive to a local hard drive. Depending on the transport model and how exposed the data is, data in transit can be less secure than data at rest.

Data in use is, as the name suggests, data that's being accessed or consumed by someone or by another application. Because data in use is subject to individual transgressions and error, it's also data in its most vulnerable state. Encryption, authentication, and access permissions play a key role in protecting data in use.

E2E stands for *end-to-end encryption*. If you're a WhatsApp user, you'll notice that it boasts end-to-end encryption as a security feature. This means that "the cryptographic keys needed to read a message are only accessible at the endpoints of the communication: the sender and the receiver, to the exclusion of intermediate parties such as service providers."[15] It's secure to only you and your reader.

ISO/IEC 27001 stands for *International Organization for Standardization (ISO)* 27001, an information security management system certification that's often required to show that a vendor or system manages data securely.

SSO stands for *single sign-on*. SSO allows a user to log in to multiple services using one login. Legal and procurement teams often require that a vendor has SSO capabilities, and you'll likely need to work with IT and the vendor to implement it.

SOC 2 means *system and organization controls (SOC)* 2. If you're involved in procuring or building a tool that will handle PII, you will likely be asked for a SOC 2 report. To get a SOC 2 report, a certified public accountant (CPA) must do an audit. The audit evaluates how well your vendor (or you if you're building the tool) securely manages personal data.

UUID stands for *universally unique identifier*. You may have an acronym that matches your organization's name. For instance, when I worked at Atlassian, we had an AUID: Atlassian unique identifier. A UUID is a 128-bit value used in software and encryption as a distinct label. It's used to present user data to people across the organization without revealing identity.

---

15 Ethyca Team, "Data Privacy Acronym List," *ethyca* (blog), June 17, 2020, last updated September 30, 2022, https://ethyca.com/data-privacy-acronym-list

Data Privacy Through Partnership

It's easy to assume that the law is black and white and that a good legal or security partner should simply be able to tell you what to do—and what *not* do. But once you start working with lawyers, you'll quickly come to understand that enabling respect in research isn't a journey of absolutes.

With the advent of the Information Age, the contexts in which people share data are rapidly changing. So rapidly that ethics, legal, and security professionals must constantly work to learn and adapt if they're to keep up with the evolution, including use cases for research. And though privacy regulations are documented in excruciating detail, it takes the skill and time of a specialist to decipher how a law might be appropriately applied within a particular context (or argued should something go wrong). To get the answers that you need, you'll need to help legal partners understand your context so that they can provide legal advice and guidance in ways that support research, participants, and the organization.

> **NOTE PRIVACY BY DESIGN**
>
> *Privacy by Design (PbD)* isn't a commonly used term, but the concepts are vital to the work of respect in research. In the foreword to a white paper called, "Privacy by Design in Law, Policy and Practice,"[16] former FTC Commissioner Pamela Jones Harbour wrote: "Accordingly, there must be some balance between regulation and innovation. One way to achieve that harmony is to embed privacy features from the beginning, starting with the design specifications of new technologies, i.e., Privacy by Design (PbD)." In the spirit of PbD and with the right partnerships, to some extent, you can co-create how legal regulations are applied to research within your organization, particularly if you partner with legal from the start (and ideally in ways that don't stunt or stutter researchers in doing their best work).

---

16 Ann Cavoukian, "Privacy by Design in Law, Policy and Practice: A White Paper for Regulators, Decision-Makers and Policy-Makers" (white paper, GPS by Design Centre, 2011), https://gpsbydesigncentre.com/portfolio-item/privacy-by-designin-law-policy-and-practicea-white-paper-for-regulatorsdecision-makers-and-policy-makers/

LEARN THE LANGUAGE OF LEGAL

Law and data privacy is a language unto its own. But you don't have to get a Harvard law degree or don a Supreme Court robe to be able to engage successfully with partners working in legal. It's helpful, though, to understand key terms so that you can follow and contribute to the conversation easily. This list is by no means exhaustive, but it's a good start.

DSAR, DSR, SAR all stand for *data subject access request*, depending on where in the world you are and the legislation you work within. Depending on your jurisdiction, you may need processes in place that enable people to make a DSAR (and you to find all the information you hold about them), which means for rigorous data handling processes. DSAR processes simultaneously support RTBF processes.

HIPAA is the Health Insurance Portability and Accountability Act of 1996. It's the United States' federal medical privacy law, which protects health information about people. The HIPAA act is important if you collect and retain health information, such as someone's accessibility status, as part of your research process. HIPAA is often relevant when research is conducted with people living with disabilities, for instance.

ISO/IEC 27001 is an industry-recognized international standard for how organizations manage digital information: financial information, intellectual property, employee data, and data handled by third parties. ISO stands for the *International Organization for Standardization*.[17] Privacy teams often require that the vendors you work with, particularly those that handle participant data, are ISO 27001 certified. There are, however, more than a dozen standards in the ISO/IEC 27000 family, all of them dealing with information security standards.

RTBF is the *right to be forgotten*. Some privacy laws, like the GDPR, CPRA, and LGPD require that you have systems in place that enable someone to request that personal information be deleted. RTBF is also sometimes called *right to erasure*.

---

17 "ISO/IEC 27001: 2022," ISO, edition 3, 2022, www.iso.org/standard/27001

Lawyers and information security partners have an unfair, and often unwarranted, reputation within the research industry of being pedantic blockers of research progress and freedom. But, unless you've happened across someone particularly unfriendly, if you take the time to understand their world and help them understand yours, you'll find that they're often fantastic (and fantastically interesting) people to work with. To make the most of these critical alliances, you should:

- Remember that the law isn't binary: it needs interpretation. So, allow enough time for legal partners to understand your context, undertake the necessary due diligence to provide legal guidance, and build out new internal templates and processes, if needed. These teams are often just as overwhelmed with requests as you are, so work with them to get on their roadmap.

- Help lawyers or security specialists learn about your research environment while you learn about theirs. This mutual respect and knowledge exchange can do wonders for the partnership (and boost your ResearchOps expertise, too).

- Help lawyers understand users' needs so that regulations can be interpreted in ways that are pragmatic for researchers, observers, consumers, participants, and research operations?

- Articulate the data that will be collected during the research workflow or a particular part of it and how it will be used and managed throughout the data governance lifecycle. This will give specialists the information they need to work confidently. A list of information to prepare is shared in, "Articulate Your Data Landscape."

- Keep partners informed of process changes that might alter the levels of legal compliance or the organization's risk profile. No one wants to get caught out, and partners always appreciate a sharp eye on the ground. You'll score brownie points!

- Work in tandem with partners to meet ethical standards that go above and beyond the basic legal requirement. Lawyers are equally invested in risk management as adherence to regulations.

Articulate Your Data Landscape

Legal partners are in demand and time is of the essence, so you'll score valuable partner points if you come to meetings prepared with answers to these defining questions. Besides, this work is also highly relevant to the work of data governance, so you'll need to know the answers. Here's the information that most legal partners care about and need to know to make good legal and risk-management decisions:

- **Data types and use:** What kind of data are you collecting, and how will you use it? Special category or personal data will need special attention and rigor in how it's protected.

- **Data retention:** How long will you need to retain the data to fulfill the purpose that it was collected for? What's the minimum data that you need to retain to achieve the research goals?

- **Data minimization:** Can you limit the data you're collecting to include only data that's essential to make sure that research is well-executed and well-utilized?

- **Data permissions:** Who has access to sets or types of data, and what could they do with it—particularly to cause harm? Is it possible to track contraventions in data access or usage? What are the risks if the wrong person got hold of the data?

- **Data access:** Do you have technical measures in place that enable you to access all the data that you hold about someone should they request it, say via a DSAR?

- **Data deletion:** Do you have protocols in place to delete the data once you no longer need it, to comply with data retention and deletion policies or in response to a RTBF request?

- **Data storage and aggregation:** Where will the data that you collect be stored? How much data will be gathered in one place about a person? The more data you collect about someone, the easier it becomes to identify them, and the higher the privacy risk. That said, sometimes companies have specific data stores or systems that are purpose-built to store personally identifiable data, so dispersing data all over the place may not be either feasible or right.

- **Transparency and fairness:** Do you openly and clearly share with someone what information you're collecting about them, and what you'll do with it and why?

- **Managing risk:** Lawyers spend a good deal of time not only analyzing and interpreting law but also determining the risk involved in doing something and how to mitigate those risks. A lawyer will often need to provide legal guidance with both the research goals and the risk tolerance of the business in mind. They will ask: Does the organization have a defensible position should something go wrong? Has it done everything reasonably possible to mitigate the risks?

PII AND SENSITIVE PERSONAL INFORMATION

PII and *sensitive personal data* are terms that you will bump into regularly—the term has already been used several times in this book.

Personally identifiable information (PII) is information that relates to an individual person. PII usually includes name, address, phone number, date of birth, signature, email address, and bank account details. But different regulations describe differently what pieces of information are considered personally identifiable. Even pseudonymized collections of data about someone can become increasingly more identifiable as they grow, so work with your expert.

Sensitive personal information or "special category data" can create significant risks to the rights and freedoms of people if abused. Because of this, handling special category data requires even greater protection, so it's especially important to work with legal partners. Sensitive personal information varies by jurisdiction but often includes:

- Race or ethnic origin
- Political opinions or membership in a political organization
- Religious beliefs and affiliations
- Philosophical beliefs
- Membership in a professional association or trade union
- Sexual preferences and orientation
- Criminal record
- Health information
- Genetic information or biometric information

Participant recruitment panels can contain sensitive personal information, which may present an increasingly more complete identifiable profile as data is collected over time. For instance, you may continue to collect information about someone via recruitment screeners. Researching with people who are living with a disability can also mean handling health information, which is considered sensitive. If you're working with sensitive personal information or enabling open-text fields in forms that give people the opportunity to share sensitive information, you'll need to allow more time (and sometimes budget) to address ethical, legal, and data governance needs.

Mind Your Scope

With the maturing of privacy regulations and the scaling of research, ResearchOps teams increasingly need to define what constitutes "research" versus other forms of customer or end-user engagement. This is particularly true in democratized research environments where the differences might be less obvious. For instance, if a product manager has an ad hoc chat with a customer, are they doing "research," or does "research" require a particular workflow? Rather than protectionism and pedantry, it has weight because it outlines the scope of research operation's work, responsibility, and risk profile in terms of ethics, privacy, and security. Should ResearchOps be responsible for the do-no-harm of *all* customer interactions and the storing of *all* data shared during customers' interactions, or only in contexts that involve part or all of the research workflow? What defines "research" specifically? It's a weighty question that you'll be forced to explore eventually. Over time, the work of ResearchOps can explode, also when it comes to participant recruitment, if you don't create a widely shared answer to this tricky question.

BUILD AN OPS TEAM

A POWERFUL TEAM OF TWO

You don't need to a hire a crowd of people to deliver respect in research. In fact, quite the opposite. Most efforts are project-based and can be delivered using a self-service operating model and with the help of automations, but you'll still need people behind the scenes to keep everything ticking over. In an ideal world, one of those people would be a full-time data governor.

Data governance is an emerging job within the professional data-driven world. A data governor should:

- Set a research data governance strategy, which should form part of your respect-in-research strategy and even your knowledge management strategy—and then deliver it.
- Build a data-wise culture that supports the ongoing safety, findability, and accuracy of data.
- Be the go-to expert for how to handle data properly.
- Make sure that data handling practices are responding to changing data laws.

- Build and maintain data pipelines, which break or change over time.
- Keep a finger on the pulse of new best practices and advances in data handling technology—in this case, specific to research.
- Align people, processes, and technology to achieve data governance goals.

In a Nutshell

While Plato was the first to systematically explore the nature of knowledge, it was his student, Aristotle, who first plumbed the depths of ethics: the study of what is morally right versus wrong.[18] Aristotle considered ethics to be a practical philosophy: He believed that ethical behavior could only result from a well-tuned internal compass, not externalized rules and theories. It's a point that's highly relevant to respect in research. ResearchOps should do more than simply lay out behavioral ideals in a code of conduct or give participants a form to sign. Instead, you should aim to make ethical and compliant practices part and parcel of how the research team rolls. You can do this by:

- **Addressing all three parts of the respect-in-research trifecta.** A chain is only as strong as the weakest link, so make sure to address ethics, data governance, and data privacy equally and synchronously.

- **Building respectful research practices into workflows.** As you map systems, consider where and how to help researchers do no harm—and come to no harm for doing research. You might use tool tips, easily-adhered-to checklists, templates, protocols, support and training, or automations.

- **Partnering with legal professionals to design (and not just align) on protocols.** The goal is not just to tick regulatory and ethical boxes, it's also to empower researchers to do their best work, and participants and observers to take part, in ways that feel good for everybody.

---

18 Richard Kraut, "Aristotle's Ethics," *The Stanford Encyclopedia of Philosophy*, Fall 2022, Edward Zalta & Uri Nodelman (eds.), https://plato.stanford.edu/archives/fall2022/entries/aristotle-ethics

CHAPTER 10

Money and Metrics

A s any statistician, mathematician, economist, analyst (or quantitative researcher, for something closer to home) will tell you, numbers are a narrative: they tell stories. Whether read as money or metrics, numbers are powerful communicators of network, culture, and connection. They offer unhindered views of who knows who, who's doing what, when, and how often. Numbers arbitrate the "bottom line" and "buy-in:" what you *do*, and *do not do*, is often ruled by numbers. They dispassionately narrate both success and failure—they can cut you down to size or build you up with good news and investment—and they can light the path for wise strategic direction, such as when to stop, sustain, or pivot your operations.

These tweet-sized stories say it all: "In Q3 FY24, 120 employees observed 40 research sessions. 20% more than last quarter. More employees are being exposed to customer needs first-hand." "Last quarter, 200 participants took part in research studies, but only 110 consent forms were registered. Who is dropping the ball, and why?" "On average, XYZ product team runs twenty research studies per quarter. They don't have a full-time researcher, nor do team members attend training sessions. How can we get them involved?" "The exclusivity deal with XYZ recruitment vendor has saved the organization $120,000 in recruitment fees within the last financial year; on average, $800 per study. Recruitment is 20% more cost efficient." The power of ten digits!

To successfully deliver research that's scalable, you'll need to see numbers as more than just pesky administration. For *money* is power, the ultimate enabler, and *metrics* are a goldmine of operational information, strategic direction, and leverage for buy-in. Both money and metrics can be complex companions, but with basic fiscal knowledge, dutiful and mindful administration, and excellent fiscal partnerships, your operations (and research, by proxy) will shine for the numerical investment.

A Highly Efficient Cost-Center

Ask any ResearchOps professional what takes up the most of their time and is least rewarding, and they're likely to say that it's administrating money (and procurement, which is related). Research requires a good deal of behind-the-scenes financial work to operate (see Figure 10.1 on page 234), a point that isn't immediately obvious, perhaps because neighboring disciplines like interface design or content strategy tend to rely on a discrete set of contractors and tools that are, by contrast, simple to administrate. But a scaled-up research practice and its operating system tend to include more than a small set of relationships and things to use. The administrative overhead commonly involves:

- Buying, building, and renewing an often-extensive kit of tools.
- Tracking, funding, and administering pay-as-you-go participant recruitment, transcription, and thank-you gift platforms (egift cards, charity, or swag options).
- Procuring and paying participant recruitment agencies for ad hoc work.
- Supporting the procurement and payment of a wide assortment of contractors and agencies for an array of research needs.
- Funding the renting or building of research labs.
- Handling the expenses of field research.
- And last, but not least, gaining and managing the investment needed to mature research operations.

But more than simply administering budgets, renewals, procurements, and transfers—the (vital) paper-pushing work of financial operations—ResearchOps should also:

- Forge partnerships with finance, procurement, and accounts.
- Get research management onboard with these team's requirements.
- Set up standardized financial workflows.
- Produce an operating budget to enable accurate financial planning, which will smooth the flow of funds to maintain operational continuity. Scaling is proportional and sustained growth, per Chapter 1, "Research Does *Not* Scale—Systems Do," so continuity is crucial.
- Produce operating reports to track spending, enable accurate forecasting (again), and show financial efficiency, impact, and growth.
- Make sure that the organization meets important financial goals and milestones.
- Enable the organization to snap up windfalls and other opportunities (where preparation and speed are key).

In less mature teams, this invisible but vital operational work is often spread among managers and researchers who look after their own needs. At small scales, this distributed setup is good enough, but as the research practice grows, your financial administration will also grow in scale, scope, and complexity. In this case, the constant fiscal learning curves that come with doing this work intermittently—every time a researcher does a study, they experience a financial learning curve—can slow down or halt entire programs of work creating frustration and wastage for everybody. (Note: The first list of tasks previously set out is distributable, while the second is less so.)

Also, if no one owns this work outright, who will look after the collective needs of the research practice? Financial colleagues tend not to have the bandwidth or knowledge (or interest) to hone the management of money so that it honors the goals of research specifically; their focus is across multiple teams, and their priority is the broader organization. Because there is *so* much power in being money savvy, and financial clout is essential to scaling, it's a no-brainer to centralize capabilities for managing and optimizing the flow of money, in and out.

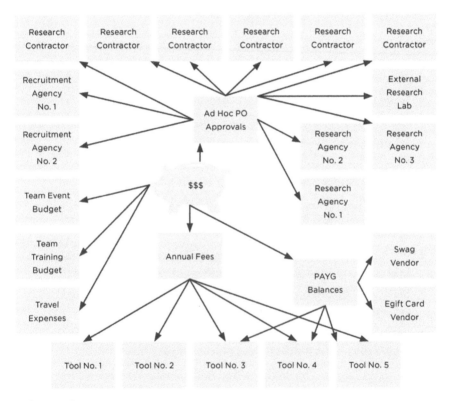

FIGURE 10.1

The ongoing administration needed to keep the financial lights on can quickly build up, especially when efforts aren't coordinated across a research practice.

HIRING A MONEY AND METRICS ADMINISTRATOR

I've had a full-time financial administrator on my team, and it's an excellent role to have. The financial responsibilities and tasks listed previously make a good job description for an energetic "numbers nerd." While you could hire an accounting or Master of Business (MBA) graduate, remember that the work is highly administrative, and you're not looking to replace or duplicate the skills of internal finance partners. Instead, hire someone entry-level who's excellent with spreadsheets and numbers, collegial, organized, a great communicator, and has a grasp of basic financial concepts and terms—or an ability to learn. And, if appropriate, involve key finance partners in the hiring rounds, because operations is nothing, if not a team sport. As ever, the role of ResearchOps is not to supplant the hard-earned skills of specialists but to create a bridge.

Getting Money for Research

The words "getting money" might conjure up images of *Shark Tank*, a reality television series in which would-be entrepreneurs nervously pitch their idea to a room full of wealthy and sharp-witted investors. While polished pitches and "working the room" are excellent skills for getting money—career-amplifying, whatever your role—they don't describe the more supportive role that ResearchOps can play in creating financial energy for research. When it comes to getting money for research (and just about anything else), there are four operational tactics that are simple but powerful.[1] They are the following:

1. Partner with finance.
2. Plan with an operating budget.
3. Make use of underspend.
4. Be a first-class spender.

---

1 The four operational tactics describe day-to-day efforts. Of course, you can also boost your chances of getting money to scale research (and ops) by aligning with the organization's strategic priorities. See "Follow the Money" at the end of this chapter.

WORDS FROM THE WORLD OF FINANCE

There are common words that will pop up in your meetings with finance partners, and it's worth having at least some sense of their meaning. Also, although these accounting fundamentals hold true for all organizations, how they're instituted can vary wildly, so check if your finance partner says something confusing and ask what they mean.

Actuals describes the actual amount of money that you've either spent or gained at a point in time (as compared to a budget, which is an estimate of what you will spend or gain). As a side note, money that's gained is also called *revenue*, but it's rare for research or ops to *directly* generate revenue for a business. Both research and operations are cost centers, so the focus tends to be more on money spent—ideally, as efficiently as possible and in line with the budget.

Accruals are income or expenses that must be included in the accounting period, although money hasn't yet exchanged hands. Some organizations have an accrual threshold, which means that costs under a certain value will not be accrued (the lump sum will be spent at the time of purchase), which can cause lumpiness in your financials.

Accounts payable (AP) refers to money that's owed by the organization to vendors. AP can also refer to the department within your organization that's responsible for making payments to vendors. Confusing!

Amortization refers to the accounting practice of spreading payments out over a fiscal period. This is especially relevant to tooling: if you sign a one-year contract with a tooling vendor for $100,000, the cost will be amortized at a rate of $25,000 per quarter to the end of the contract. This means that $25,000 per quarter will be allocated from your budget, as opposed to

Partner with Finance

The first and most important thing you can do in managing (and getting) money is to forge strong partnerships with your colleagues in finance, accounts, and procurement. Which is not to say that collegial connections will make money grow on trees, but it helps significantly to treat people who work in finance as more than just banks, blockers, number crunchers, and contract sticklers! Instead, see these partners as number-loving folk who want to do more than just administer your world. Help them feel involved and informed. I once invited a finance partner to visit an in-progress research lab I was

$100,000 at the time of signing the contract. It's an important point to understand when it comes to annual budget planning but note that the rules of amortization differ from company to company.

Carryover means funds that were approved to be spent in a previous fiscal year can be carried into the following fiscal year.

Financial year (FY) describes the 12-month period designated as the operating year for financial management. When a financial year starts will differ from country to country.

H1 and H2 describe the first and second half (six months each) of the financial year.

Q1, Q2, Q3, Q4 describe the quarters (three months each) that make up the financial year.

Master services agreement (MSA) is a contract between the organization and a vendor that defines the terms and conditions of all the work that will be delivered during the contract period. If an MSA is in place, you should only need to raise a PO to process a new order with a vendor.

Purchase order (PO) is a procurement document that details the approval for the purchase of services or products: when they will be delivered, by whom, and how much you will pay for them. When you procure something, you'll be asked to "raise a PO."

Prepayments define expenses that are paid in full in one accounting period that will be consumed in a future accounting period. Organizations typically have prepayment thresholds. For instance, services under $100,000 can be prepaid, but services over $100,000 must be accrued, as defined at the top of this list.

building—it was barely a lab at that point—to share the vision and express the impact of his support. This small (and genuine) gesture did wonders for our financial success across multiple programs of work and made it all the more enjoyable.

Building camaraderie will enable greater support and flexibility when needed for all parties: for example, the rearrangement of budgets to meet changing priorities; patience with mistakes; reminders to meet milestones; help getting genuinely urgent procurements in place (use ASAP rarely); and a skilled hand in navigating complicated calculations and spreadsheets. Additionally, being proactive and engaged means that financial partners will tend to come straight to ResearchOps if they have a question, which, on occasion, can reveal useful information about research requests and projects going on across the organization, or when money has become unexpectedly available.

Not everyone will be open to forming a partnership—I can think of one finance partner who was just plain mean—but, whatever the reception, share your purpose and learn their language, cadences, and needs.

Plan with an Operating Budget

If there's one skill that will help in managing money and metrics, it's mastery of that most vital of financial tools: the spreadsheet (unless you have access to a more sophisticated tool, of course). In the world of finance and operations, the most impactful thing you can do with a spreadsheet is create an *operating budget*: an accurate-as-possible financial plan for the upcoming accounting period, whether it's the quarter or the year. An operating budget will help you:

- Plan how much money you'll need for what you need, when you'll need it, and for how long. It will help make sure that the research practice isn't caught empty handed when there's work to be done.

- Understand non-negotiable operating expenses (OpEx), like participant recruitment. This budgetary heads-up can also give you a nudge to adjust ops downstream if your budget is more restrictive than you'd like.

- Ask for money when budgets are being decided—usually during your organization's formal cadence of financial planning. If you miss these important moments in time, you run the risk of being strapped for cash when you need it.

- Track your actual spend versus your planned spend throughout the fiscal period, and plan for over- or underspend. Neither are good in the world of organizational finance; you want to be just about bang on the money every time. In fact, it's common for organizations to allow a 2% tolerance for both under- and over-spend. As part of financial tracking, make sure to communicate any discrepancies with your finance partner and leader, too.

- Understand your spending trends: where and how money's being spent quarter-on-quarter and year-on-year. This will give you excellent metrics for budget forecasting, the trickiest and most important of financial tasks, particularly when it comes to variable operating expenses (OpEx).

OPEX, CAPEX, FIXED, AND VARIABLE

In preparing a budget, it's worth understanding the concepts of OpEx, CapEx, fixed, and variable costs. These financial concepts offer a useful framework for understanding the various types of spending, and how to plan and manage each. In the accounting world, expense types can have complex implications, but to manage research operations your understanding need only be top-level.

Operating expenses (OpEx) are the ongoing day-to-day expenses needed to keep things operating, i.e., KTLO. Research OpEx often includes staffing costs (contractors, researchers, and ops), software license fees, participant recruitment costs, transcription costs, thank-you gifts, and travel costs. The bulk of a research organization's expenses tends to fall under OpEx.

Capital expenses (CapEx) are one-off investments typically spent on large tangible assets, which will give the organization a lasting benefit, usually for longer than one tax year, at least. A research lab, a large suite of research field kits, or the production of licensed and reusable research training assets are all examples of spending that is CapEx.

Fixed costs are expenses that don't change over time. Fixed costs tend to be outlined in contracts or cost schedules and remain the same for the life of the agreement, unless both parties reach a new agreement. The most common research fixed costs are software-as-a-service (SaaS) tools, staffing, or contracts where a fixed price is preagreed. In other words, it's easy to know what you'll spend because it's written in a contract.

Variable costs change over time, depending on how much you use. Participant recruitment, thank-you gifts, and transcription services are all examples of variable OpEx costs. While the fee may stay the same (a $40 admin fee for each research participant recruited), the amount of service you use (the number of participants you recruit) will vary depending on demand, which will alter how much you spend. Variable costs are the hardest costs to forecast, but it's worth every bit of effort to get close.

If you regularly track your usage metrics and actuals via standardized operating budgets and reports and keep an eye on figures that foretell scales in demand, your variable OpEx costs will become much more predictable. But it takes time to develop this kind of rigor, which is why measurement should be constant throughout the phases of operational maturity. Most maturing operations teams must initially use finger-in-the-air "forecasting" until they've mastered their operations metrics and can forecast accurately. If you've got no metrics or previous financial data to lean on, put your best guess forward confidently. A best guess ask for money is better than no ask at all!

Depending on your finance partners' bandwidth and the nature of your partnership, you may need to do more or less of your budget planning solo. At the end of the day, they'll need to understand and approve your budget, so always ask partners for support—equally, aim to be their easiest partner! For instance, if they have a budget template or tool that they like to work with, use that. Also, the end of each quarter and the financial year are typically extremely busy for finance, so it's generally appreciated if you're prepared ahead of the rush. Finally, finance partners will often have records of your operating expenditures over several quarters and years, which, particularly if your operations are young and your metrics shallow, can help significantly with forecasting which will help their work, too.

To develop an operating budget, you should:

1. Gather previous spend (actuals) and usage metrics to forecast future potential expenditure (budget), particularly in the case of variable OpEx costs.

2. Review current operations and ops roadmaps to understand what you need to build, specialize, or scale within the financial year, and what you'll need to fund it. For instance, will you need more seats for XYZ tool to meet a growing demand?

3. Find information about expected organizational growth for the budget period. (HR might have this information.) Will there be an uptick in the number of new hires whom you'll need to support? If so, budget for it.

4. Using these data points, lay out a forecast of your total expected spend. Ideally, you'll present these as two separate budgets: OpEx and CapEx.

5. Note which expenses are fixed and which are variable. In other words, determine costs that are dependable (fixed) and costs that may adjust (variable) throughout the fiscal period. And make sure to track your actual spend on variable costs regularly—monthly is good—so that you can utilize underspend, and manage overspend, wisely.

6. Finally, create a financial calendar to track vendor renewals and terminations. This will help you prepare for contract negotiations, which can mean budgetary adjustments up, down, or out, in the case of a contract ending.

Good annual planning will put you in the driver's seat, and yet, as vital as an operating budget is, and it is *vital*, it's not the full score to determine financial success in operations. When it comes to getting investment for new research and operations initiatives, you'll need to do more than just master partnerships, budgets, and spreadsheets. You'll need to learn how to scoop up money at opportune moments.

Make Use of Underspend

If you've ever lived in London—perhaps this happens in all big cities—you may have noticed a sudden flurry of maintenance descend on the streets just before the end of the financial year. Spending money on street repairs is a time-worn method for dispensing of excess cash quickly, which is exactly what councils need to do to reduce their *underspend*. If you don't use it, you lose it, in the world of institutional finance! There are two morals to this story: First, make sure that you spend the money you ask for on schedule. If you don't, you won't get it back next year. Second, if you're well-prepared and known to be a reliable spender, at particular times of the year, you can often meaningfully make use of others' underspend, and, because you'll balance their books, too, finance will love you for it.

How underspend is managed differs from organization to organization and even from country to country, so it's worth understanding your organization's financial cadences and rules inside out, and then keeping in touch with finance partners about potential places to help out. Finally, have two or three ready-to-go projects lined up that might usefully utilize underspend in your budgetary department—*your* underspend, too, if needed. In the past, I've hired a content strategist to improve onboarding and support materials using money that other people were too disorganized to spend. I've also delivered the pilot of a video repository tool, a project that delivered a ton of learning.

Be a First-Class Spender

The reason that underspend exists is because it takes effort, planning, and good execution to spend money, and not everyone does this well. If you don't spend your allocated budget, you're not guaranteed to get it again the following accounting period, and, on a more anecdotal note, your finance partners are less likely to fight in your financial corner for funds in the first place. Spending well means that you've spent your money *on time* and *in line* with the original budget and agreement. To spend money well:

- Make sure that you've got a solid plan and the resources in place to execute before asking for money. This sounds simple, but it's amazing how often this trips things up. The ResearchOps Planning Matrix will help you get this right. (See Chapter 4, "Planning Realistic ResearchOps.")

- Track your *actual* spend versus your *planned* spend as set out in your operating budget. This will help you adjust to accommodate over- and underspend.

- Acknowledge that things will go wrong, and mistakes will happen. But, if you track your money well, you can usually get ahead of a problem by communicating with finance.

- Consider the cadence of teams like procurement and finance who will often need to support you to spend money by approving contracts. Finance and procurement teams are often under the whip as they wrap up the financial year-end, so, again, try to be ahead of the game.

- Last, but not least, negotiate *fair* contracts with vendors. Prices are often negotiable, but don't overdo it.

NEGOTIATE FAIRLY WITH VENDORS

It's good financial practice to secure good deals for your organization with vendors. While it's tempting to wrangle deals on your own, if the deal is more substantial, it's often worth asking procurement partners to take over. A good procurement partner will be an arch negotiator and an impressive force in the room, so let them do their work. Besides, letting someone else handle negotiations will help you maintain a more collaborative relationship with the vendor in the long-term, which is important.

At this point in history, many research tools are up-and-coming startups, some more established than others. Depending on how desperate they are to have your organization's name (and money) on their books, they may offer you an excellent deal to inaugurate the relationship. If a deal looks too good to be true, question it with the vendor and procurement. It takes a significant amount of work to onboard and sustain a vendor, as shared in Chapter 8, "Tactical Tooling," so it's worth making sure that the financial relationship is sound and sustainable for both parties in the long term. I've been known to bargain a vendor up, and it paid off! The relationship lasted five years and counting.

Making the Most of Metrics

Money is a kind of metric, but there's so much more "numbers power" to be had in operations beyond counting dollars and cents. Brought together as an operating report or dashboard, metrics—money and stats—can reveal the story of potential efficiencies, trends, drop-off, and churn (just as they do in product management) and offer powerful insight for how to hone research culture, craft, and connection. *This* is the fuel that ops should use to inform strategic direction, as opposed to just administer things on request. Every mature research operations team should deliver a like-clockwork quarterly and annual operating report. By consolidating and analyzing metrics, you can:

- Track the impact of both research and your operations. What's working, not working, and what has space for improvement.

- Spot financial trends. Metrics are essential in forecasting variable OpEx costs as part of budget planning. Over time, you learn that, on average, you spend say $20,000 per month on participant recruitment. Armed with this information, during the annual budget planning cycle, you place a request for $300,000 to cover the financial year's costs for participant recruitment (including 25% more to accommodate growth). Perhaps you also notice that you spend more in Q1 than any other quarter. In preparation for Q1, you bolster your participant recruitment platform's accounts to accommodate the uptick.

- Track drop-off across your operations pipeline. Perhaps 200 sets of research videos were uploaded during the quarter, but only 100 consent forms were signed in the same period. Why are people not practicing informed consent, or are they simply not using your systems?

- Understand who's using your systems and services, how they're using them, and how much. If you're supporting democratized research, is a particular team regularly doing a lot of research? Perhaps it's time to see if they need more research support, or might they be open to hiring a full-time researcher?

- Tell the story of your return on investment, quantify your impact, and bring qualitative understanding to your spending. As per the earlier example, metrics are useful in showing your impact and building measured (and measurable) cases for investment.

- Empower yourself to answer questions about usage, growth, or spend at a moment's notice, which sounds trivial, but it's genuinely empowering! When I managed a ResearchOps team, I was regularly asked for "a win in a sentence" to share with executives, and, thanks to operating metrics, I was rarely lost for good content.

Perhaps you're a researcher or product manager and you're thinking: "but none of this is new" or "quantitative data offers only half the view; it's not the full picture." And you're right. Using metrics to measure success, failure, and impact is far from new, and it's not the full view. But it's rare to see this standard product practice properly applied in the emerging field of ResearchOps, and certainly not as operating reports. To offer a full-fleshed narrative of the user experience and your operational impact, metrics should be balanced with qualitative insight, story, and color. Go mixed-methods research! Though a mass of detail might be left out of view, an operating report comprised of nugget-sized numbers stories is essential to gaining investment for both ops and research, and quantifying (and selling) impact, too.

> **NOTE** **OPS REPORTS AND DASHBOARDS**
>
> To track and present your metrics, you should produce an operating report or a dashboard—or both. An operating report tends to be a more readable asset for stakeholders (use it to tell short-as-possible stories with action points), while a dashboard will help the operations team keep track of numbers, and write up reports, more readily.

Metrics Tips and Tricks

Collecting, managing, and readying metrics is painstaking work that makes up an analyst's full-time job, so unless you are in fact an analyst, acknowledge any valid feelings of imposter syndrome and keep on because this work is every bit worth the effort. There are a few challenges to be aware of, but, put aside the need for perfection, and they can all be overcome:

- First, not all vendors provide usage metrics by default, so you may need to work with each vendor to get metrics automatically or manually (or accept that you can't reliably get them at all). Include this enquiry when you're onboarding a vendor or designing a new service, as shared in Chapter 8 and within the ResearchOps Planning Matrix.

- Manual data may be inaccurate, incomplete, or late, and sometimes it can change, which can frustrate the idea of a standardized and timely report, but so be it. Better late than never, and better done imperfectly than dumped!

- Your metrics will often come from multiple places, which means that the format of your data will vary wildly. It can be brain-busting work to consolidate it into a meaningful story, especially if you don't have a degree in statistics or a good grasp of mathematics. Do the best that you can.

- It's worth noting again that to deliver value, an operating report needs at least one year's worth of consistent tracking to be able to see trends. In the meantime, you'll still get value from understanding how people are engaging, or not, with your operations. When you're wondering whether the investment is worth it, it is, so keep going.

- It's unrealistic, in most cases, to think that all research interactions will happen within the confines of your operations. However well-laid your paths, people will tend to find their own way to recruit participants or say thank you for taking part in research. It's worth acknowledging up-front that your reporting is confined to what you can measure and be okay with that.

There are simple ways to surmount these challenges, and they are the following:

- **Measure what matters.** The multiplicity of vendors and tools used by most ResearchOps teams, and the current lack of data interoperability between them, means that creating a consolidated dashboard for the full operational pipeline, or a full complement of research workflows, is extremely difficult, if not impossible, and not always an efficient use of time. Instead, devise a dashboard, a spreadsheet will do, that tracks the metrics that will be most useful (this might take some experimenting) and which you can consistently get your hands on.

- **Data continuity is essential—so is simplicity.** Make sure to standardize your metrics, reports, and dashboards, and keep them simple. Standardization will allow you to compare figures across quarters, and eventually years, and to spot trends and forecast supply, demand, and budgets. It's far more important that your report is regular, accurate, and consistent than a show of your newfound (or well-honed) data analyst skills!

- **Make reports actionable—*then* act.** Aim to produce an operating report both quarterly and annually to offer a consolidated fiscal-year view. But remember that creating a report isn't enough; you need to act on the insights, too. Diligently creating reports is one thing, but if you fail to act because you're onto the next thing or simply too busy, they're of little value. To remedy this, when the report is hot off the press, walk team members and stakeholders through the highlights and agree on time-bound actions. I like to use SMART goals—specific, measurable, achievable, relevant, time-bound—to define short-term tasks that need to be done. Finally, the first years' worth of operating reports may feel a little light on reward, but if you're consistent and patient, by the second year, you'll start reaping the rewards in spades.

But What Metrics?

Your research operating system is likely comprised of a plethora of research tools, assets, services, and platforms. Each of these assets should, by default or design, have usage metrics to contribute to your metrics pool. The next question is, "What metrics should you gather?" It's always a good idea to track the things that senior leadership are interested in,[2] but there are baseline metrics that you should track, too. For example:

- **Usage data:** Monitor the number of users or "bums in seats" of both vendor-delivered and in-house research tools and how much they use the tool. You might discover that a tool isn't being used or that you have too many seats—or not enough.

- **Engagement metrics on research content:** Research reports, guidance, blog posts, playbooks, and more. Tracking research reports, for instance, can inform trends in people's interests and where the research practice might pay extra attention.

- **Participant recruitment metrics:** Number of participants who joined a research panel; number of participants who were recruited; number of researchers who recruited participants using the panel; types of participants who were recruited; no shows; and panel diversity metrics.

---

2 Executives are often interested in seeing one consistently reported metric expressed in a simple sentence that aligns with their priorities or values. This makes it easy to remember and use.

- **Research training attendance:** Completion, and drop-out, or people who simply didn't show up—an unfortunate metric, but it's sadly often true.

- **Onboarding metrics:** Number of people who attended or completed an onboarding pathway versus new hires you'd have expected to take part (success metrics).

- **Requests for support and problems solved:** The number of support requests you received; average response time; average time for a request to be resolved; and satisfaction scores. Is your support service effective or not? Support is crucial but unglamorous work that can be hard to fund. Metrics are invaluable in making support services more efficient and the extent of the work (and impact) more visible (see Chapter 7, "Seamless Support").

- **Financial metrics:** How much you spent on what, and in what time frame. You should be able to gather this information in detail from your finance partner.

If you're the kind of person who reads an entire book, or watches a movie until the last credit, even if you knew that it wasn't for you in the first chapter or scene, then this note is for you: experiment with collecting a metric but, if it doesn't pay off over time, say a year or two, drop it from the roster. It is not worth the cognitive noise and time.

Follow the Money

U.S. President Joe Biden is quoted as saying: "Don't tell me what you value, show me your budget, and I'll tell you what you value." I don't tend to quote politicians—someone is always offended, so forgive me if this time that's you—but the quote is just too bang-on to leave out. In other words, if you want to know where interests truly lie in an organization, follow the money. And if you want money, attach yourself to an existing priority (or sell your idea so that it becomes a priority).

One of the biggest challenges in getting buy-in for ResearchOps is that it's emergent: a new practice with little documented history, till now. It takes effort and time to help people understand what ResearchOps is and why it's worth investing in, particularly in a world where people are often still digesting the value of research. A secondary challenge is that good operations are almost invisible: it's behind-the-scenes work which, like Wi-Fi, is only noticed when it's not working well. Depending on your context, you may need to pitch, prioritize, and trumpet your heart out to get more investment. But there are a few tricks for navigating these challenges, like using metrics to make your impact tangible, partnering with others to create a bigger voice, utilizing underspend, or attaching yourself to existing priorities.

Ideally, priorities are enshrined in the form of strategies (organization-wise, stakeholders, and research), but if that's not so, look to objective key results (OKRs) or key performance indicators (KPIs), or whatever form of goal-setting your organization uses. For an even more direct route, per Biden's quote, ask your finance partner who's getting the money and why. You don't always want to follow the money, but attaching your work to existing priorities is a powerful way to gain much-needed investment to grow your impact where it matters most.

JUDO: LEVERAGE EXISTING ENERGY

The Japanese martial art, Judo, holds the motto "maximum efficiency with minimum effort," which can be translated as using the energy that's already before you. It's a motto that's as useful in martial arts and life as it is in gaining buy-in for research and ops.

Several years ago, I arrived at an organization with set ideas as to what my remit should be: to deliver "research operations," as per my specific definition at the time. The executive in charge expressed interest, instead, in an initiative that was operational in nature and related to research, but which sat outside my immediate remit. I made the case to focus on "pure" ResearchOps, and the initiative was given to someone else. While I battled for funding, the executive's initiative (no surprise here) had no such problem. While there's merit in maintaining focus on the core intent of your work, it pays to stay open-minded to priorities that might use your skills alternatively. While I did deliver extensive operations value, I might have enhanced the capacity and clout of ResearchOps more broadly and significantly if I'd been open to more immediate (and well-funded) opportunities.

In a Nutshell

For some people, money and numbers, not to mention spreadsheets, can be daunting. But they don't have to be. If you can bring your numerical world down to earth with strong fiscal partnerships and an operating budget and report, you'll be empowered to run your operations reliably, grow the impact of both research and ResearchOps incrementally, and snap up last-minute opportunities. These are the points to remember:

- **It pays (literally) to have a great relationship with finance partners.** Aim to give them just enough context to make the interaction meaningful—not so much as to overwhelm—and, wherever possible, adopt their ways of working to ease their cognitive load. If you can get ahead of the game when it comes to budget planning, you'll shorten their to-do list at extra busy times of the financial year, which will keep you in their good books.

- **Be an excellent spender.** Spend according to a budget that's driven by operating metrics (good forecasting) and strategy (good planning) and keep your finance partner in the loop if there are hiccups or surprises, good and bad, that might impact the budget.

- **Find ways to scoop up underspend within your department or team.** It helps to have one or two projects preplanned and ready to go (an MSA in place, with only the need for a PO approval), and then let financial partners know that you have the capacity to mop up underspend in ways that are valuable.

- **Consistently monitor metrics using a dashboard and present highlights and action points in short-as-possible quarterly and annual operating reports.** A dashboard will help ops get a quick overview of the latest numbers, while a report can tell the story of important narratives and actions. But make sure that action points make it onto the right people's to-do lists and get the attention they deserve.

CHAPTER 11

Getting Priorities Straight

Enter "prioritization" in a Google search, and you'll be delivered countless lists about how to achieve focus, be more productive, stave off distraction (or procrastination), and achieve your goals. Whether the focus is life hacking or business management, recommendations are fairly standard: create a list, identify what's important, consider effort versus impact, be realistic about scope, and learn how and when to say no—all good advice that can be applied to prioritizing research as much as anything else. In other words, you can create a research intake form, a prioritization matrix, models for assessing feasibility and scope, and a research roadmap, to echo innumerable research blog posts. But there's a catch. While they're all infinitely practical assets, and you'll learn more about them right here, they're *not* the full answer to managing research prioritization so that it's scalable across more (and more complex) scopes, relationships, and roles.

The research practices that need to invest in operations tend to include dozens of researchers, hundreds of people who do research, a large and complex network of stakeholders, and countless variations in study needs and types. Not to mention myriad expectations as to what research is and what it does.

In its simplest form, getting priorities straight is about tempering researcher overwhelm and delivering a modicum of order and calm, which is good for everyone. But prioritization is also the primordial soup of research strategy, per Chapter 2 "Lost and Won on Strategy," and deciding what research *does* get done and does *not* get done, in other words, how to respond to demand, should inform both researchers' and operations' focus.

More than just managing demand, getting priorities straight should be about *culturing* demand, which is the most potent string to the bow. Instead of just bringing control to a call-and-response operating model, prioritization systems should methodically empower research leaders to harness hard-won demand top-down: to purposefully power, direct, and shape how people see and experience research and, therefore, the quality of the requests received.

Driven by Demand

For decades, research pioneers have battled to create demand for research, a battle that's largely been won—even if many disciplines assume that good results can be achieved by DIY, which isn't always the case. The opportunity now resides in how demand is crafted and managed, how roadmaps (and relationships) are worked out, and whether researchers thrive, survive, or nosedive as demand for research, if not always researchers, rapidly ramps up. The most common reactions to this tricky, if triumphant, spot in the scaling-research journey are to:

- Hire more researchers (assuming the investment is available; there never seem to be enough).

- Deploy researchers strategically to focus on certain topics or teams, leaving others without (a prioritization in and of itself).

- Democratize research[1] (adopt the often-extensive task of helping other disciplines do research well—or well enough—and engage with research assets and experiences).

- Develop prioritization workflows and assets (to capture and cull long lists of research requests down).

- Or all of the above, if you have the privilege of resourcing and choice.

These are all fair tactics, each with a unique set of pros and cons. But for any of them to be truly effective, the *relational* attributes that backdrop prioritization (the "contracts" you agree to with people, knowingly or not) must first be set right. Deciding what's important and what should be done is a position that's hard won, so prioritization is often all bound up with perceptions, reputations, reporting lines, party lines, who owns what, and other complicated, context-dependent, and sometimes messy human stuff. While it's on researchers and managers to shape relationships one-on-one—not everything can or should be operationalized—ResearchOps can provide a positive backdrop that researchers and stakeholders can build on. For prioritization is as much about how you manage the *source* of the demand as it is about managing the *demand* itself.

---

1 When there's a call to democratize research, it's often because nonresearch disciplines are already independently doing research. If this is the case, operations can deliver significant benefits—financial efficiencies, and better ethics, and compliance and research outcomes—by centralizing operations for things that are already being done ad hoc.

The Scourge of Unspoken Contracts

These days, it's not unusual for research teams to balloon in size seemingly overnight. (Unfortunately, they can also halve in size when the economic environment gets tough.) In many cases, depending on the team structure, this growth is fueled by individual stakeholders, say a content strategy or design leader, from across the organization funding research head count to support their own research demand. While the growth of the research team is great—it's done a world of good in amplifying the profession of research—the "contracts" that govern the exchange are rarely properly fleshed out. Instead, they're often built on conversation, assumption, and collegial trust. That's not to say that colleagues shouldn't be trusted or that every interaction should be formally ironed out, but that inconsistency, time, and change are rust to important things not preagreed or written down.

In the case of embedded researchers, a modus operandi in which researchers are dedicated to a particular team, it's an easy leap for a stakeholder to assume that in acquiring or funding the head count for that researcher, they've bought the right to decide what *their* researcher does. And why not? But if the embedded researcher is centrally managed, the research manager might feel that it's their job (and skillset) to manage *their* researcher's work. And again, perhaps the researcher feels it's their place to prioritize their own roadmap and backlog.

The critical question is: Who's in charge? At small scales, and depending on personalities, this awkward triangulation is often easily ironed out. But as the research team grows, it can quickly become onerous, or simply disharmonious, to manage, negotiate, and renegotiate the countless partnerships that underpin the success of research. Not to mention the professional experience of those mired in the tussle for power. Because to own prioritization is power.

However, your researchers are deployed—embedded, centralized, working agency-style—ResearchOps can help set "subterranean stabilizers" to enable smooth and productive partnerships above ground. Specifically, ResearchOps can:

- **Drive engagement with research.** Help people know that you exist and who you are via an *engagement strategy*.
- **Clarify how to team up.** Help people know how to work with you via a *collaboration model*.

- **Collect, prioritize, and track.** Standardize assets like a research intake form, prioritization matrix, checklists, and research roadmaps and backlogs.
- **Bring in the PgMT troops.** Hire and empower project or program managers to help straighten priorities out.

At first glance, several of these tactics might seem oblique to prioritization, but they're a powerful quartet in shaping collaboration and prioritization dynamics for the benefit of everyone.

Deliver an Engagement Strategy

As someone who works in the research profession, you likely know that we all make assumptions based on perceptions, and they're called *biases*. Aware of it or not, we engage with people and the work they produce based on conscious and subconscious perceptions about who we think they are: Are they an expert or an upstart, respected or relegated, trustworthy or suspect, similar or different from us, higher up the "food chain" or not? Your stakeholders' biases, assumptions, and past experiences about research, your research team, and even the research profession at large will impact their expectations about research and how they should work with you. But as any marketing, brand, communications, or business whiz will tell you, you needn't leave any of this to chance. From simple to complex and consistent to one-off, there's a lot you can do to help people engage with research in ways that will support its success in the long term. You can shape people's perceptions about:

- Who you are: your team, values, and mission.
- Your impact and expertise.
- What *good* research looks like, what it takes to deliver, and the value it brings.

In Chapter 2, I wrote about the importance of amplifying the value of research, which is often hidden from wider view within distributed efforts and relationships. Engagement efforts should improve the efficiency and success of distributed relationships, but they should also amplify and shape perceptions about the value that the research practice is delivering. There are countless ways to help people engage with and develop a particular understanding of research and the value it provides. The sky is really the limit. Implementing even simple engagement tactics can quickly add up to a full-time job, not to mention too much noise for your audience, so it's a wise move to choose what to do, and what *not* to do, based on specific engagement opportunities and challenges. In other words, you should develop an engagement strategy.

An engagement strategy comes from the world of marketing and defines a communications plan for shaping how people perceive something. In this case, "research" as a craft within the organization and "research" as a discrete team. An engagement strategy should also map the touch points that you currently have, and think you should have, with your target audiences and how you'll make those moments of interaction educative, purposeful, and impactful.

Depending on your touchpoints, goals, context, and culture, you might deliver one or two low-maintenance and one-off investments like the following:

- **A brand identity:** Logo, typography, and other visual assets to use on all research communications and assets, including research reports.

- **A well-designed website or intranet:** Express your brand and communicate who you are and how to work with you. You might communicate your collaboration model here, too. (Note: Make sure that engagement content chimes well with support content. See Chapter 7, "Seamless Support.")

- **Opportunities to watch research first-hand:** Enable stakeholders and others to observe research sessions live or attend research share backs or "watch parties," a tactic that supports research impact via tacit knowledge sharing, too.

- **Team swag:** High-quality stickers, t-shirts, or hoodies to deliver a bit of fun and pizzazz!

If you've got access to additional skills and resources in the long term, you could choose to deliver higher-maintenance tactics where an ongoing effort is necessary, and consistency is key. You might decide to:

- Publish regular communications, such as newsletters, announcements, blog posts, and posters to help people know that you exist, what you offer, and how to engage with you.

- Host research events such as expert talks, user story sessions, or research town halls and invite everyone to join.

- Be proactive about helping researchers (and research ops) speak at organization-wide events, like all-hands meetings or other team meetings.
- Use organization-wide or team-specific communications channels to regularly engage target audiences.

> **NOTE DOUBLE DOWN IN REMOTE CONTEXTS**
> If your company works remotely, you'll need to work doubly hard to make sure that research efforts, whether events, reports, and even the existence of a dedicated research team, aren't missed, forgotten, or overlooked.

Engaging people with research, and changing their perceptions about it, can also be achieved with less obvious tactics. Perhaps your goal is to help research stakeholders understand first-hand what well-executed research looks like, so they can make better judgments about the research they come across (and trust researchers' skills and opinions). To help achieve this goal, you might encourage people to attend research sessions moderated by a skilled researcher—not to empathize with the customer, though that's got obvious benefits, but to reset their benchmark for excellence.

The Subtle Art of Perception Shifting

Research observation is commonly called *exposure hours* per Jared Spool.[2] The original intent of exposure hours, and it still stands today, is to help team members "see it to believe it" and to understand end-users' needs and experiences firsthand. But there's a secondary benefit that's useful to leverage: watching skillfully moderated research also exposes people to high-quality research, which can tacitly change how people engage. Under research leader Leisa Reichelt's leadership, the UK's Government Digital Service (GDS) used both exposure hours and training to help make "user research a team sport," and it worked like a charm.[3] Viewed through the lens of engagement, research training also tacitly teaches the investment and

---

2 Jared Spool, "Fast Path to a Great UX—Increased Exposure Hours," *Center Centre—UIE* (blog), March 30, 2011, https://articles.uie.com/user_exposure_hours

3 Tingting Zhao and Steph Marsh, "How We Use Training to Help Make User Research a Team Sport," *User Research in Government* (blog), October 15, 2019, https://userresearch.blog.gov.uk/2019/10/15/how-we-use-training-to-help-make-user-research-a-team-sport/

benefit of doing research well, and, therefore, the value of investing in researchers who are purposefully skilled. Training isn't only an opportunity to enable others to do research themselves.

Engagement efforts aren't an overnight or single-pronged transformation. The potential to shift research perceptions, and the type and quality of research demand, is a power that should not be underestimated in straightening priorities out.

ONBOARD AN ENGAGEMENT EXPERT

The communication needs of a scaling research team can quickly blow out into a demanding full-time job. If you can, hiring a communications or public relations expert won't go to waste. There's a trick, though, in how you manage this new and largely unexplored role. While engagement efforts tend to start out creative and innovative (the build/buy and standardize phase), they can quickly become heavy on logistics as the focus shifts toward keeping successful efforts afloat (KTLO). To satisfy the full remit of this role, you'll need to hire someone who's content to take on both the creative and administrative elements of engagement, and then work with them to keep the balance right.

If you can swing it, you'll hire two people: one to handle the ongoing creative operational work, which can include other creative work like marketing for participant recruitment, the building of talent pipelines, and so on. Then hire a second person to handle the administrative side of things, like scheduling events. If you don't get this balance right, you'll frustrate your creative person and fail to make the most of their skills.

Clarify How to Team Up

A strong engagement strategy will help invest a sense of research identity within your organization. But if researchers are externally funded, as they so often are, it's crucial to formulate clear guidelines around collaboration so that everyone understands what it means to purchase or partner with research and, crucially, who's in charge. A collaboration model will take time and investment up front, but it's a one-off and low-maintenance investment that will support the success and impact of research in the long term.

A collaboration model is a modus operandi for how a group of people should work together. It should document the roles, responsibilities, resources, tasks, information sharing, commitments, duration, process, and planning, and what to do if things go wrong, in particular circumstances. For instance, a collaboration model for hiring an embedded researcher could define:

- How to hire a researcher.
- Who manages the researcher?
- How much time will a researcher give you?
- Who should offer performance feedback?
- Who should be involved in decision-making, including prioritization?
- Who gets to make the final decision?
- What to expect from a researcher: the resources they'll have access to, where they'll need collaboration and support, and what they'll be able to produce in what time, in general.

> **NOTE** **INSPIRATION FROM THE WORLD OF BUSINESS MANAGEMENT**
>
> A collaboration model takes inspiration from the world of business management where it is called an *engagement model*. Engagement models tend to pertain to external relationships and typically take the form of a legally binding contract, a level of formality that isn't usually necessary in the context of in-house research.

Depending on your context, team structure, and research strategy, you could standardize collaboration models for:

- Embedded researchers
- Researchers who work agency-style
- Research agencies and contractors
- People who do research, such as designers or product managers, and how you can (or can't) support them
- And, not to forget, how to work with research operations

A collaboration model must be agreeable to everyone involved—this work isn't about being doctrinaire or shutting down one-on-one relationship building—and it should be flexible. You can work with researchers, leaders, and stakeholders to formulate a basic and standardized model (or a set of models) for how things are done, which you can use for years to come. The model should be mutually beneficial, in most cases, most of the time, and something that researchers and stakeholders can equally believe in and build on. Make sure to revisit it on a regular cadence to make sure that it remains applicable and is being used. Finally, document your collaboration model in ways that fit your organization's culture and communication style.

Collect, Prioritize, and Track Research

One of the chief complaints that researchers have about their profession is the constant feeling of being bombarded by requests for more research, all of it a priority, all of it needed ASAP. Engagement and collaboration efforts can help settle some of that noise, but how research requests are collected, prioritized, and tracked will deliver additional sanity for everybody. And there's another benefit: By creating a dashboard of research requests, you'll gain a valuable overview of the forces driving research demand, which is invaluable in shaping the strategies of both research and research operations. In honor of this effort, ResearchOps can standardize these areas:

- **Research request form:** How to collect requests.
- **Prioritization framework:** How to work priorities out.
- **A research study tracker and backlog:** How to track commitments and delivery over time.

A Research Request Form

A research request form, sometimes called an *intake form*, gives people the opportunity to let you know that they think they need research and to describe what they want to learn as a result. Request forms are common across the wider research profession. For example, scientific, civic, and academic research disciplines all make use of request forms to manage research demand. Disciplines like visual design use request forms to manage their workloads, too, so this advice is nothing new. A centralized process for managing requests can also provide research leadership and operations with an overview of the needs and gaps in terms of research capability and knowledge—a valuable set of metrics for illustrating demand and the need for specific investment. But if you want these benefits, you'll need to be considerate about the tool you use to power your intake form. Also, if you have a properly staffed research library, you could respond to request forms with a list of existing reports that are relevant to the topic (along with an invitation to explore the request further, if appropriate).

TOOLS FOR SCALABLE RESEARCH REQUESTS

It's common to use a survey tool to provide an intake form, which works okay. But to enable intake forms that are efficient and scalable, you're better off using platforms like Zendesk, HelpDesk, or Jira Service Management. These kinds of platforms offer form-filling, triaging, and response-tracking capabilities, which will enable you to manage and track requests more easily across the prioritization workflow and use collected data to understand your research demand better over time.

Help desks also give you the capability to set up a centralized support desk, as shared in Chapter 7. This can empower researchers to access requests relevant to them while also maintaining a centralized "super-admin" view of *all* research request queues. This is important because it will enable you to:

- Offer research management, including program management (PgMT), a consolidated view of request types: what, when, and who.

- Easily hand over requests if someone's on leave, or if there's a staffing change.

- Enable knowledge management efforts such as a referencing service or literary reviews, as mentioned earlier, to flesh out research requests in the queue—not every research study needs to be brand new.

- Access a valuable birds-eye view of recurrent themes, knowledge gaps, and capability needs for the near-term or future. This is an invaluable set of metrics in guiding the future state of a fit-for-purpose research team.

Provided it's easy to find and fill out, a request form can save researchers hours of time in meetings, messaging, and questioning, and it can support inclusive and fair prioritization, too. Work with your researchers and stakeholders to design (and test) a request form that balances form-filling with one-on-one conversation. Here are a few "should" and "should-nots" to help:

- A request form should help requestors articulate what they want to learn from research, the decisions they hope to make as a result, how critical their request is, and their expected timeframe.

- Avoid asking stakeholders whether they want a survey, diary study, or usability test, for instance. Stakeholders often don't have the skills to decide on the best course of action for answering a research question and asking them to think about it will only invest them unnecessarily in a particular methodology. Besides, it's the researcher's job to work out the details.

- Your request form shouldn't be so onerous to fill out that it kills the conversation on opening. Keep it focused on the high-level information you'll need to triage the request and have a meaningful one-on-one conversation.

A Prioritization Framework

While prioritization is still relatively basic within the user research profession, scientific, civic, and academic communities have research prioritization down to a fine art. These long-established research disciplines have done extensive work on operationalizing research prioritization, so take inspiration, because you needn't navel gaze in solving the time-worn challenge of too much demand—or the wrong kind of demand.

A particularly interesting example comes from NASA's scientific research arm. Even as far back as 1995, scientists experienced similar prioritization pains to user researchers: specifically, research prioritization was being done by someone else! If you've heard a researcher bemoan this scenario, a quote from NASA's 1995 report, *Managing the Space Sciences*[4] might make you smile: "The scientific community is uncomfortable with priority setting in which factors beyond pure scientific merit come into play, partly because it implies judgment by people who are not experts." The author goes on to say: "If scientists are to play an expanded role in establishing priorities, they must engage in deliberations outside their areas of primary scientific expertise. They must also become accustomed to the intrusion of nonscientific [sic] factors." And so it is all too often the case in the world of user research!

---

4   Space Studies Board, National Research Council, *Managing the Space Sciences* (Washington D.C.: National Academy Press, 1995), https://ntrs.nasa.gov/api/citations/19960013913/downloads/19960013913.pdf

Prioritization in user research requires that researchers build knowledge around both business-specific areas and wider organizational themes and build skills of collaboration and negotiation—key research competencies. As you already know, an engagement strategy and collaboration model can help iron out relational complexity and ambiguity, and an intake form can help collect and record requests for research, but the key question remains: How should those requests be prioritized and, critically, *by whom*?

To answer this, it's useful for researchers to have a well-adopted and accepted prioritization framework. In its fullest sense, a prioritization framework is a consistent practice and criteria for helping people make unbiased, timely, and strategic decisions about what to do next. It should be designed in collaboration with researchers, but it should also make sense to collaborators across other disciplines—it's a social and cognitive framework for decision-making, so it will fail if buy-in isn't broad and inclusive. As such, prioritization frameworks can get complex and philosophical, which means that the work can easily get stuck in the weeds. So, if you're driving the effort, aim for progress over perfection and cross-functional cohesion: your framework should be intrinsically useful for everybody. There are three key parts to a prioritization framework:

- A prioritization workflow
- A set of prioritization criteria
- A prioritization toolkit

A Prioritization Workflow

A prioritization workflow encapsulates general guidelines and protocols for how research requests should be prioritized: who does what, what tools to use, and triggers for certain actions. In other words, it's logic. A good prioritization workflow should help make sure that the research organization is spending time and energy where it will have the most impact, the overall direction of which is guided by the research strategy, per Chapter 2, and that prioritized projects are set up for success. That last point is critical. I've seen research contractors hired to attend to a particular brief, only to discover that the tooling, participant recruitment capabilities, and compliance needed to do their work wasn't available. Assumptions were made, and no one thought to check if the team had the necessary capabilities. I've heard of researchers making overly enthusiastic promises and managers double-booking researchers, too! It's these kinds of wrinkles that a prioritization workflow (and framework) can help iron out before it's too late.

In essence, a prioritization workflow is an operating model for prioritization, so it could be self-service, full-service, or a hybrid of both. For instance, ResearchOps might:

- Provide researchers with the training and tools they need to independently work out priorities with their stakeholders (a self-service operating model).

- Alternatively, ops might centralize elements of the workflow by triaging requests and then assigning appropriate researchers to follow through on prioritization.

- A full-service model could include deploying research program managers (Research PgMT) to run prioritization sessions in collaboration with researchers and stakeholders, assuming that you've got that kind of capacity.

A workflow can also define a trigger. If your operations include a self-service research library (or better yet, a team of reference librarians), you might insist that each new research request triggers a past research review to understand what's already known about a topic, thereby making for better-informed prioritization. The sky is the limit as to what your workflow could define and do. Whatever you do, make sure that the system is contextual and simple to use.

A Set of Prioritization Criteria

Prioritization criteria are central to the success of any prioritization framework. Criteria are a set of fundamental standards or principles that should be used to measure projects in the process of prioritization. The most commonly used criterion is effort versus impact, but criteria can and should be more extensive. NASA's previously mentioned document, *Managing the Space Sciences*,[5] offers an excellent example of prioritization criteria, which you might take inspiration from or simply steal. Their criteria are:

- Merit
- Potential payoff

---

5   Space Studies Board, *Managing the Space Sciences.*

- Novelty
- Originality
- Importance
- Timeliness
- Methodology
- Uncertainties
- Conceptual reservations
- Importance in relation to other projects
- Feasibility and cost
- Policy
- Organizational strategy
- Ethics

A Prioritization Toolkit

After you have a discrete set of criteria with which to vet a potential study, consider the tools you'll need to enable researchers and their stakeholders to match projects against criteria in ways that ease out biases and stimulate productive stakeholder conversations. It's common to use a 2x2 matrix to help plot what should be done and *not* done (see Figure 11.1). While a 2x2 matrix is useful, it's worth remembering that a 2x2 only enables comparisons across two variables, like effort versus impact. Comparing two variables *is* a valuable exercise—a matrix is also easy to understand in a prioritization workshop—but you'll still need to plot other important variables such as feasibility, risk, timeliness, and ethics.

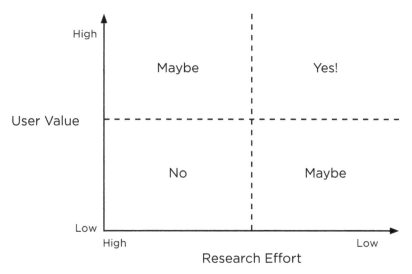

FIGURE 11.1

2x2 matrices always (and only) compare two variables. In this example inspired by an NN/g article,[6] the variables are "user value" versus "research effort."

It's worth recalling the scenario shared earlier of the contractor whose time was wasted because none of the essential capabilities were available to support their work: while the research impact might have been worth the effort (effort versus impact), the project wasn't feasible to deliver and shouldn't have been executed. In short, a 2x2 matrix isn't the be-all and end-all of research prioritization. As a result, your prioritization toolkit should include several tools for assessing whether a research study should be given the green light. You might include the following:

- 2x2 matrix to understand where a project sits in the context of two key criteria. This can be used to drive productive conversation in a stakeholder meeting or workshop.
- Feasibility checklist to understand whether what's important or desired is also achievable from a time, resourcing, and budgetary point of view.

---

6   Sarah Gibbons, "Using Prioritization Matrices to Inform UX Decisions," *Nielsen Norman Group* (blog), May 27, 2018, https://www.nngroup.com/articles/prioritization-matrices/

- Ethics checklist to make sure that a project that is desirable and feasible is also ethical. For more on ethics, see Chapter 9, "Respect in Research."
- Any other tools that researchers need to be able to run prioritization sessions and document outcomes. It's highly likely that these are already provided by the organization, and you won't need to lift an operational finger.

RECOMMENDED READING

TOOLS FROM THE FIELD

The uptick in research demand means there's been an uptick in the number of blog posts about research prioritization, too. Here's a sample of the some of the best:

- "Building a Framework for Prioritizing User Research" by Jeanette Fuccella: **https://uxdesign.cc/building-a-framework-for-prioritizing-user-research-ed46622ead99**
- "Using Prioritization Matrices to Inform UX Decisions" by Sarah Gibbons: **www.nngroup.com/articles/prioritization-matrices/**
- "NCredible Research: Positioning the Power and Potential of Research Using a Reflection and Roadmapping Framework" by twig+fish **https://medium.com/twigandfish/ncredible-research-positioning-the-power-and-potential-of-research-using-a-reflection-and-5373adc371c0**
- "Researching the Right Thing Versus Researching the Thing Right" by Katarina Bagherian: **https://dovetailapp.com/blog/researching-right-thing-versus-researching-thing-right/**
- "5 Prioritization Methods in UX Roadmapping" by Sarah Gibbons: **www.nngroup.com/articles/prioritization-methods/**

A Research Study Tracker and Backlog

Once a project has been prioritized, its progress should be tracked as part of a research study tracker (and if the study is put on ice, it should become part of a regularly groomed research backlog). Especially as the research organization grows, it's increasingly more important to run a consolidated study tracker—a single source of truth—that tracks what's been agreed to, what needs to get done, and what didn't get done and why. A study tracker can help:

- Researchers and their stakeholders know what's coming up next and what they've agreed to do.

- Researchers and their stakeholders understand capacity during prioritization, which can trigger reprioritization, too.
- Research leaders track focus and impact across research studies, which can inform strategic decision-making.
- Operations to plan resources, capacity, and budget *ahead of time*!
- Research knowledge managers understand what they can expect to land in the research library once the study is done.
- Show gaps in research knowledge and offer insight as to possible primary and secondary research themes.
- Managers and researchers track their output and workload. It's often sensitive and inaccurate to use this data as a dry measure of a researcher's performance, but it can be useful in guiding performance conversations or preparing for promotions later on.
- Provide useful metrics and stories about research supply and demand—plus scope and impact.

NOTE **TOOLS FOR TRACKING WORK OVER TIME**

Every organization has a different way of tracking work that's been planned, completed, or backlogged. While there are countless tracking or road-mapping tools from which to choose, it's often best to adopt the tool that's already widely used across the organization. By toeing the line, you'll save yourself the unnecessary hassle of onboarding another tool, and you'll be instantly integrated with the workflow (and language) of your partners, which will support easy collaboration, too.

In a multilevel research organization, you'll also need to decide whether you'll centrally track studies per researcher, per team, or across the entire research organization (centralizing tracking). Or perhaps you'll simply provide researchers with standardized tools and templates to track their own work (decentralized tracking). Again, this depends on your size, capability, and prioritization workflow, but if you decentralize research tracking, remember that you'll lose out on valuable and consolidated operating metrics.

Bring in the PgMT Troops

It's not uncommon for researchers to find themselves playing monkey in the middle. For example, their research manager has asked them to say "no" to a particular research project, but playing hard ball with the team of people they work with isn't easy, particularly as they need to build a connection and trust to succeed. If the research organization has the capacity to hire program managers to help drive conversations around priorities across the research team, this can help researchers build strong and trusting research relationships while PgMT (program management) say "no" where necessary. Not all researchers want this work taken off their hands—some would gasp at the suggestion—but in many scaled-up contexts, it's a welcome and necessary relief that can release a researcher to focus on what they do best: research.

Project Management and PgMT Are Not Operations

Operations is often conflated with project management or program management (PgMT), but the disciplines are not synonymous (see Table 11.1). The work of ops is to deliver the infrastructure that enables the research organization to work effectively, which naturally involves project and program management as *every* act of delivery does. There's nothing to stop you from managing a program management team as part of operations. Actually, it might be a great way to grow your impact—just don't conflate the two disciplines. Too many teams say they're delivering "operations" when they're really program managing, which does the field a disservice. To make the distinction clear, here's a comparison between project management, program management, and operations management.

I've long dreamt of hiring a program manager not to do PgMT but to design and deliver the systems that enable strong PgMT across the research pipeline—the exact tactics that have been shared in this chapter. If you find yourself in the fortunate position of being able to live that dream, give your PgMT expert this chapter and then let them get to work. There's enough work laid out in this chapter to keep a visionary PgMT with a flair for operations and service design usefully busy for years.

TABLE 11.1   THE DIFFERENCE BETWEEN PGMT AND OPS

Project Management	Program Management	Operations Management
Delivers a single and discrete project that has a defined start and end. It tends to be short-term work.	Delivers multiple projects as part of a broader strategy. This work tends to be long term.	Delivers and maintains operational efforts in the "forever term." Once you've got good operations in place, the work never ends.
Coordinates day-to-day work and tracks the success of a project.	Implements strategies, defines success metrics, and manages collaboration more generally and across multiple projects.	Designs, delivers, and maintains efforts and infrastructure that enables the organization to add value and deliver its strategies.
Works as part of a temporary team or embedded in a single team in the long term.	Works as part of an established team but can also contract. Might also manage project managers.	Works as part of an established operations team and often manages a team of operations experts.
Work is constrained to a specific time, cost, and scope.	Part of an annual planning cycle that focuses on funding discrete pieces of work.	Part of an annual planning cycle and must fund both new projects and ongoing operational costs.
Measured by the success of individual projects or teams, including compliance with the budget and timeline.	Measured by the success of program strategies and organizational objectives.	Measured by the success of operational impact, continuity, and organizational efficiency in the long term.

In a Nutshell

To get research priorities straight in ways that are efficient, effective, and scalable, you'll need to do more than just devise an intake form and a 2x2 prioritization matrix, although they're helpful assets to have. Instead, you'll need to take a more people and systems-oriented approach (again, sociotechnical systems) that includes the following:

- **Shape people's perceptions and biases about research and your team via an engagement strategy.** Communications that let stakeholders know who you are—your team, values, and mission—and how to work with you.

- **Clarify how to team up using collaboration models that are standardized to particular contexts.** A model might define how to hire a permanent researcher onto a product team, and how everyone should work together in the long run.

- **Put systems in place that help researchers and stakeholders to request, track, and prioritize research efforts—and operations and management to get a 10,000-foot view of where efforts are being spent.** This usually includes a prioritization workflow and criteria, and a toolkit that includes a study tracker and backlog, all of which can help researchers know how to manage their workload and leaders to know with pinpoint accuracy where investments are going.

Making the Most of Research Talent

S caling research, scaling any discipline really, requires an investment in people. Until now, at least within this book, efforts toward scaling research have focused on tactics like actively managing research knowledge and logistics or forging a democratized research environment. All primary operational work. But researchers are by far the most important asset and enabler in maturing and scaling research, and it's *their* experience that will help bring stability and skill (and character!) to the research team's efforts. As a research team scales, you must scale more than just tooling, training, and scope. How researchers are hired, managed, and enabled to grow—how they feel about their team and their work—is central to successful operations. Research operations should empower research leaders to manifest a team that is appropriately skilled, motivated, inclusive, and diverse, which doesn't happen on its own. You'll need to enable meaningful conversations between managers and direct reports about performance and growth. And you'll need a strong talent pipeline to keep pace with the demand for head count, an increasingly common (and glorious) problem to have. Of course, that's putting aside the spate of layoffs in 2022–24.

It's here, at the crux of people's personal and professional lives that ResearchOps can do some of its most important and foundational work. But "important" needn't mean complex, overwhelming, or in need of extensive investment. In fact, this is the least complex and demanding of the eight elements of ResearchOps, an irony you should take full advantage of. One enthusiastic mid-to-senior-level person with support from leadership and human resources (HR) is all it takes to bring this pivotal element to scalable life: to craft and standardize HR's efforts in honor of the goals of research, and in honor of the researchers (and ResearchOps professionals, not to forget) who will make it all thrive.

Know Your People Partners

To make the most of research talent, you'll need to forge a close partnership with HR, otherwise called *human resources management* (HRM) or the *people team*, a friendlier name for the same work. HR manages how staff are recruited, managed, rewarded, and retained within the organization and across the people management pipeline (see Figure 12.1).

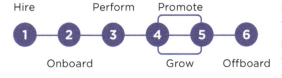

Hire   Perform   Promote

① — ② — ③ — ④ — ⑤ — ⑥

   Onboard   Grow   Offboard

FIGURE 12.1

The people management pipeline includes six phases from hiring to managing performance and offboarding.

Each phase of the people management pipeline is related to a team of HR specialists. It's worth understanding their unique roles and scope and, by proxy, the phases of people management, so that you can specialize each phase toward the goals of research.

- **Talent acquisition (TA):** Is responsible for managing how the company acquires talent. This job includes building a strong employer brand and talent pipelines, sourcing applicants, maintaining relationships with high-potential candidates for future roles, and supporting managers in hiring well.

- **Talent management (TM):** Focuses on how talent is managed once they're onboard. TM delivers employee training programs, enables and supports performance management, and defines and standardizes how employees are coached, mentored, promoted, and rewarded.

- **Human resources business partners (HRBP) or people partners:** Tend to work with leadership and, ideally, operations to specialize HR efforts in support of the team they're serving. ResearchOps and the team's HRBP should be close allies.

If you don't have access to an HRBP, there's still a lot you can do independently to make the most of research talent. But first, there's a discrete piece of work that is critical, and it's foundational to everything else: a competency framework for research. To create one, assuming that you haven't got one in place already, you'll need HR's approval and help.

> **NOTE** **PULL THIS WORK UP THE PRIORITY LIST**
>
> You might spot that much of the guidance offered in this chapter could be applied to any other discipline, and you'd be right. You can apply it to design, marketing, or operations, and you'll deliver positive impact, for sure. But the wild ride of scaling and maturing research, along with the intensive need and focus of the remaining seven elements of ResearchOps, means that these important people-focused efforts are often missed. Make them a strategic priority, though, and when the research practice starts to rapidly scale again, you'll be well ahead of the game.

MAKING THE MOST OF RESEARCH TALENT   275

A Research Competency Framework

A competency framework is a people management asset that benchmarks the behaviors, responsibilities, and skills of a specific discipline (job family), role, and level. If you were to search for "competencies" online, you would find as many variances in the precise framing of "competencies" and "competency frameworks" as there are colors in a rainbow, but the intent is always the same: to identify tangible traits that a team member should demonstrate on the job. There are three key aspects to a competency framework:

- **Behaviors:** Define *how you do* a job.
- **Responsibilities:** Define what you *need to do* to fulfill a job.
- **Skills:** Define what you *need to know* to do a job.

Without a clear and shared competency framework specific to research in place, it can be hard—impossible, even—for both researchers and their managers to gauge and manage excellence (and tardiness) in performance, maintain motivation, and talk about promotions and growth. It's important to note, that competency frameworks shouldn't be used as a checklist—a restrictive rubric for scoring performance and growth, which is a point that human resources will reliably reinforce. Instead, it should be used as a tool to guide conversations about performance and growth between managers and researchers, and managers and managers during performance cycles. In fact, more than just guiding conversations about performance, a competency framework should be used to inform communications throughout the people pipeline (see Figure 12.2). It should help:

- Define job descriptions during hiring.
- Frame interview questions and templates for unbiased hiring.
- Set role expectations for both managers and new hires during onboarding.
- Drive performance conversations and feedback.
- Facilitate equitable growth and promotions.
- Make sure that practices across the people pipeline support diversity and are equitable and inclusive (DEI).

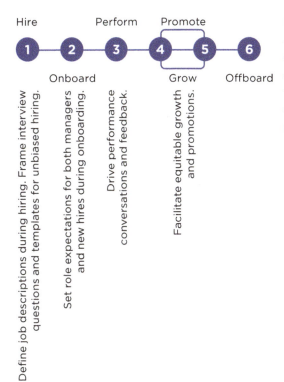

FIGURE 12.2
A competency framework should be used to inform activities throughout the people pipeline, not only during the annual performance cycle.

Competency framing is specialized and sensitive work bound up with precedence, job families, salaries, HR systems and, of course, *people*. It's important to acknowledge that in changing or making tangible the expectations of someone's day-to-day work, you may also alter team members' professional (and personal) experience for better or worse. Make sure to work sensitively *and* work with specialists. Do the work once and do it well. And finally, advocate to bring people—leadership, managers, and researchers—on the journey so that you have buy-in from day one.

Expect that you'll need to develop at least two frameworks: an individual contributor (IC) framework and a framework for research management. Even if you don't yet have a management level, a management level is part and parcel of scaling. So, if you're already knee-deep in the work and have HR's attention, produce all the frameworks at once. ResearchOps professionals are equally desperate for professional shape and direction, so don't forget to look after them, too!

Attract the Right Research Talent

With an HR-approved competency framework in place, you'll have delivered a powerful and lasting contribution to scaling research, and you'll have set a solid foundation for additional tactics that ops can deliver to augment HR's centralized work. The first of these efforts comes in attracting the right research talent to apply for open roles. Depending on your context—where you are, who you are, who you're looking to hire, and the state of the economy—you may find that competition for experienced researchers is steep.

To become a talent magnet and get ahead of the pack, you'll need to offer a compelling compensation and benefits package, for sure. But well before the money talk begins—a task that's allocated to hiring managers, finance, and your HRBP to handle—operations can help get the right talent in the door. People need to know that you're looking to hire, and that you're an excellent place to further a career in research. If you work for a well-known company with a great reputation, your work here might be minimal, but every scaling research team needs a robust talent pipeline, to use talent acquisition parlance. To attract excellent research talent, ops can:

- Build a team brand.
- Encourage referrals.
- Take part in talent events.
- Make the most of graduate programs.

Build a Team Brand

Whether you own it or not, your research team will develop a reputation and a brand, and its standing within the profession, or lack of it, will impact the kind of people you can attract. If your team culture isn't sterling and word has gotten around, you've got much harder and more important work to do—work that's beyond the scope of this book. But if it's simply that you're not publicly celebrating your culture, talent, and work, read on.

There are simple tactics you can use to build a team brand, many of which will synchronously support the growth of an excellent team culture and bolster current researchers' profiles, too. It's a double win! You don't need to do all these tactics all of the time or all at once. In fact, they're time-consuming, so be strategic about what you choose to do. You could:

- Start a research blog and publish posts consistently.
- Fund and publish research that has broad public appeal.
- Sponsor well-chosen research conferences and then make the most of the event. Do more than just send them your logo. For instance, you could generate conversation at the event with a research-oriented game.
- Support researchers in speaking at professional events—big, small, and tiny.
- Run a public research event regularly. You might even invite talent that you're keen on hiring to take part or speak.
- Create resources that further the research profession globally and then share them openly.

It takes time to develop a genuine reputation that people can get excited about and trust. Plus, results don't happen overnight, so be patient and consistent in your efforts and specific about what you want to achieve (build out a communications strategy that defines how you'll attract particular types of talent), and plan ahead. Finally, if the research organization is set to scale, start your comms engines early.

Genuine contributions to the development of the research profession go a long way toward building a team reputation and brand. Here are some of my favorite examples:

- Global design company, IDEO, has an excellent blog: **www.ideo.com/blog**
- Another lovely contribution from IDEO is *The Little Book of Design Research Ethics*, which was shared in Chapter 9, "Respect in Research." It's beautifully written and, first published in 2015, has longevity. They have also released a second edition: **www.ideo.com/journal/a-new-edition-of-the-little-book-of-design-research-ethics**
- The design software company, Figma, regularly publishes beautifully designed reports—of course they're beautiful—about product design trends, analysis, and best practices: **www.figma.com/reports/**
- The UK's Government Digital Service (GDS) user research blog became a source of inspiration for researchers around the world. Research thought-leader Leisa Reichelt and I inaugurated the blog in 2014, so I'm blowing my own horn to some degree! **https://userresearch.blog.gov.uk**

Encourage Referrals

When it comes to low-effort opportunities for attracting great research talent, none fits the bill better than encouraging referrals. Research tends to be a tight-knit community, so it's likely that incumbent researchers will be able to offer unrivaled support in recruiting new talent. Referral schemes are standard practice in most companies, but they only work if people know that there are jobs on offer and that there's an opportunity to gain a bonus for a successful referral. In this case, the task for operations is simple:

- Set up a consistent communications channel to let researchers know about new roles early: perhaps set a regular email or team message about hiring opportunities.
- Remind researchers each time that the referral scheme exists and how to use it.
- Thank researchers for their help in hiring: beyond referral bonuses, find ways to acknowledge the use of their personal network.

Take Part in Talent Events

If you've managed to form a good partnership with Talent Acquisition (TA), you might be invited to take part in, or know to ask about, organization-wide events aimed at attracting new talent. Not all events will be right for research, but if you spot one that fits, get involved with enthusiasm. Or perhaps work with TA to shape one that's more suited to research and its neighboring disciplines.

BUILD AN OPS TEAM

HIRE AN EXCELLENT COMMUNICATOR

Branding, referrals, and events aren't complicated efforts, but they do require dedicated attention and a particular kind of talent. To cover this work, you'd do well to hire an entry- or mid-level person with communications, event organization, and public relations–type talents. I've had someone work in this role and would hire it again without thinking. Apart from owning this work and making the most of research talent, a communications expert can also own internal engagement for research (a team presence, emblem, and communications strategy) as shared in Chapter 11, "Getting Priorities Straight." Plus, they could own the communications strategies needed for research support and participant recruitment if they have the time. In short, this hire will not be bored for a moment.

Make the Most of Graduate Programs

Many organizations have a graduate program: an allocation of entry-level roles held for people (usually recent university graduates) who have a degree in the right area but no on-the-job experience. Graduates can be an excellent route for bringing new high-potential talent into the team, provided the research organization and managers have the time and capabilities to offer the mentorship, training, and support that's expected as part of the deal. If your organization has a graduate scheme, it's a good idea to hook into the program, but make sure that you're able to meet all the needs first.

RESEARCHOPS ALSO NEED GRADS

Getting head count to support operations can be grueling, so take advantage of graduate schemes where you can. It's easy to assume that graduates might be ideal to cover administrative jobs, but a graduate program is supposed to catapult someone in their chosen career, not hold them down. So, make sure to give graduates weighty projects that make the most of their specialist skills. In general, you'll be interested in graduates with a degree in:

- Systems and information design (to support the backbone of ops—systems—and research knowledge management).
- Public relations or communications (as per the work set out in this chapter).
- Marketing, data analytics, and engineering (to support participant recruitment and data governance).
- Finances (to support money and metrics).

Build a Diverse, Equitable, and Inclusive Team

Building your team brand and tapping into referrals, TA events, and graduate programs will give you the best choice of *whom* to hire, but *how* you hire (and how a diversity of people experience your team) is also of significant importance. There's a lot of work to be done in enabling diversity, equity, and inclusion (DEI) across the research profession, never mind within the wider industry. It's not within the scope of this book to enable you to meet these important goals, but there are things you can and must do to enable people from all walks of life to feel welcome, that they belong, and that they're valued. There are four primary opportunities, noting that this list isn't comprehensive, to improve diversity within your research team, at least at the critical point of hiring:

- Make special efforts to reach diverse populations when advertising roles both internally and externally.
- Ensure that language and images used in job descriptions are inclusive. For example, **datapeople.io** is a useful tool for writing inclusive and easy-to-read job descriptions.

D&I, DEI, AND DE&I: BUT WHAT DOES IT MEAN?

Conversations about DEI can be daunting. If you don't identify with being from an underrepresented community, you might feel that you're tripping over your words. To enable DEI within your organization, it's important to become informed and comfortable with the language and needs of DEI, just as you might with other language-sensitive topics like accessibility.

Variably called *DEI (diversity, equity, inclusion)*, *D&I (diversity and inclusion)*, or *DE&I (diversity, equity, and inclusion)*, the words can get confusing. So, what do they all mean?

Diversity means that the workforce is diverse—it includes a variety of team members from across culture, race, religion, age, gender, sexual orientation, and disabilities.

Equity, or equitable practices, makes sure that processes and programs are fair, impartial, and provide equal opportunity for everybody.

Inclusion, or inclusive practices, aim to make sure that everyone's included and feels a sense of belonging.

- Design bias out of résumé reviews and interviews.
- Make sure that all onboarding materials and work tools are accessible to employees living with a disability. It's inherent on operations to push vendors to meet Web Content Accessibility Guidelines (WCAG),[1] at least.

Many organizations now have a DEI team, and they're well-equipped for this sensitive work. If you have such a team, get to know them, hook into initiatives they're already running, and, if possible, work with them to pioneer DEI initiatives specific to research.

---

1   "Making the Web Accessible," W3C Web Accessibility Initiative (WAI), www.w3.org/WAI/

DE-BIAS HIRING

There are four common types of biases that, conscious or not, can creep into hiring:

Affinity bias describes feeling a connection with people who are similar to you.

Perception bias describes the assumptions and stereotypes you might make about particular groups of people.

Confirmation bias happens when you look to confirm existing opinions and ideas about something or a group of people.

Halo effect means projecting positive traits onto people without actually knowing them.

Bias in the hiring process can be mitigated using the foundation of a competency framework and the development of interview templates for each repeatable role: researcher, senior researcher, research manager, and so on, or operations manager, leader, coordinator, and administrator, for example. Most hiring platforms now offer features that support unbiased hiring, and some like Applied and Datapeople, a platform that can be used to write inclusive job descriptions, have unbiased hiring as their core mission. Applied has produced an extensive and beautiful bundle of guides, toolkits, examples, and templates for de-biasing hiring, which you can access for free.

Applied's "Debiased Hiring Bundle" of resources is useful: www.beapplied.com/changing-recruitment/diversity-inclusion-de-bias-your-hiring-guide

There's one more type of diversity that's important to consider in terms of talent management, too. Although hiring for a diversity of experience and skills is not one of the seven official types of diversity (culture, race, religion, age, gender, sexual orientation, disability), diversity in the skills you welcome into your team can bring strength and creativity and help avoid an echo chamber, which can stunt professional growth for researchers. Learning how to spot people who have talent but who haven't walked the standard professional line is a skill in and of itself. In the still-forming profession of research operations, it's rare to find people who match your job description precisely, so being able to spot unique talent is an essential skill.

Make Onboarding More Than a Checklist

A 2022 *Harvard Business Review* article[2] shared: "According to Gallup's onboarding report,[3] employees who have a positive onboarding experience are almost three times as likely to feel prepared and supported in their role, boosting their confidence and improving their ability to perform their role well." Onboarding is the bridge between promises made during the hiring process and a new hire's first on-the-ground experience, and it can be anything from overwhelming and daunting to exciting and empowering. To some extent, whether it's one or the other hinges on the new-hire's temperament (I tend to be more anxious in new situations), but a good deal of the experience rests on the state of your onboarding support and materials. So, make them sterling.

In addition to the basics offered by HR's talent management, there are pragmatic efforts that ResearchOps can take to help managers set researchers (and ops) new hires up for success and put out a warm welcome. First, it's important to note that a new hire will usually need to onboard at various levels, which isn't often considered. They'll need to be shown the ropes for:

- The wider organization
- The research team
- Their team
- Their manager
- How to get research done
- Their area of research focus
- Their stakeholders

Your organization likely has a standardized onboarding program to look after the first need, but it won't look after onboarding specific to research, and it certainly won't address unique onboarding needs such as how to get research done, stakeholders, and research focus

---

2   Sinazo Sibisi and Gys Kappers, "Onboarding Can Make or Break a New Hire's Experience," *Harvard Business Review*, April 5, 2022, www.hbr.org/2022/04/onboarding-can-make-or-break-a-new-hires-experience

3   *Create an Exceptional Onboarding Journey for Your New Employees*, Gallup, Inc., 2019, www.gallup.com/workplace/247076/onboarding-new-employees-perspective-paper.aspx?thank-you-report-form=1

areas. Ideally, operations should put standardized and self-service assets in place to support managers and new hires in navigating each onboarding level. You might choose to standardize:

- **Communications for announcing that someone has been hired, when they'll arrive, and what they'll be doing.** This will help keep people informed, help them welcome a new hire, and avoid unnecessary and sometimes uncomfortable surprises.

- **A checklist for managers to use in making sure they've covered the onboarding basics.** Have they set up weekly one-on-one meetings? Have they talked them through the competency framework? Have they added them to organization-wide communications channels and team meetings?

- **An onboarding checklist for new hires to guide them and set expectations around onboarding.** Out the gate, a new hire should be set up with specific tasks to get them going—meaningfully—for at least the first week.

- **An onboarding pathway that introduces new researchers to the suite of research tools, training, and support that they'll need to access while working.** Ideally, your pathway will be digital and self-paced so that it's scalable and doesn't block researchers from getting on with things immediately.

- **Easy access to a support channel or informed "buddy" to answer endless questions.** New hires should hopefully feel that there's no such thing as a silly question!

- **A social mechanism for helping (even shy) people to be social.** Help new hires build networks and social connections, especially outside of their immediate team. And *especially* if you work remotely.

- **A handy glossary of the organization's homegrown language to smooth track the new hire's collegial communications.** It can be daunting when you're new and you can't keep track of what everyone is talking about. What phrases, acronyms, and jargon should they know?

- **And, finally, a resource that few organizations have the maturity to offer but it's the ideal!** A secondary research package that covers the researcher's proposed areas of focus so they can quickly get up to speed and speak with their stakeholders confidently.

That said, whatever you produce, be aware that already bombarded new hires will feel the pressure to do and absorb everything, which can quickly result in being overwhelmed, even if that wasn't the intention. Try to keep onboarding high level and minimal, and pace onboarding so that it offers the right information at the right time and at the right fidelity. It's not unusual for an onboarding program to span 90 days, so use that time to onboard someone incrementally. In this case, the tortoise always beats the hare.[4]

> **NOTE** A SEAMLESS SUPPORT JOURNEY
>
> Onboarding crosses over closely with research support as covered in Chapter 7, "Seamless Support." When you design research support, keep the previous list of onboarding ideas handy. The experience of company onboarding, team onboarding, and resource onboarding should be unified.

Help Researchers Thrive on the Job

It takes significant commitment, effort, and investment to hire and onboard someone onto a team, not just for the organization but for the new hire, too. While a good on-the-job experience often comes down to good management, team culture, and whether people see eye-to-eye—subjective experiences that can't be operationalized—there's a lot that research operations can do to help researchers thrive and grow in their career. The two most potent strategies are:

- Make it easy to learn.
- Help researchers be seen and heard.

Make It Easy to Learn

A researcher once said to me: "We're put on a research hamster wheel, so you barely have a moment to think about your career trajectory. Promotions happen to you, or they don't, and there's little time to think about opportunities for learning." It's not unusual to hear that researchers and ops folk are stretched to the limit and career-growth seems a distant priority.

---

4   "The Hare and the Tortoise" is a fable credited to Aesop, a slave and story-teller who lived in ancient Greece between 620 and 564 BCE. In it, the slow and steady tortoise beats the hare at a race because the hare raced ahead but then decided to take a nap. https://read.gov/aesop/025.html

Over and above the learning opportunities that the wider organization might offer—an annual education budget, generalized training courses, and so on—ResearchOps can help busy researchers continue to grow by providing easy-to-attend learning opportunities, which, as an important sidenote, can have double-power because they also bring researchers together. Each of these tactics takes a fair bit of work to deliver, so choose one or two that will have impact and staying power:

- Create a research mentorship program. Both the mentor and mentee will grow in experience.

- Facilitate the organization of coaching and training opportunities for researchers, or research operations. A few years ago, I arranged a ten-week service design training for my team, and it helped us grow our skills, bond, and learn together, and shape a shared language for our work.

- Deliver an internal speaker series. Inviting thought-leaders from across the industry to speak can help researchers be inspired and look outside of their bubble.

- Help researchers go to industry conferences, even as a team. Conference attendance can also build a strong team brand and personal networks, both of which are good for hiring.

Help Researchers Be Seen and Heard

It can be daunting to write and publish a blog post, or to stand up on stage to give a speech. But writing and speaking are essential research skills and honing them in front of an audience can be rewarding and motivating. Helping researchers become strong public communicators can strengthen their motivation, skills, and confidence. Also, a strong public presence can strengthen the building of a lively team brand and your talent pipeline. Of course, helping researchers be seen and heard will also increase their chances of receiving new job offers and opportunities, but protectionism is no way to make the most of talent! This work can easily become a full-service offering and intensive, but there are scalable options that you can deliver. You might:

- Provide a list of conferences along with a link to call-for-speaker applications. You might even offer mentorship in honing applications, too.

- Offer a speaking course to support would-be speakers.

- Similarly, offer a writing course to support would-be writers.
- Publish guidance or offer support for navigating the approvals that someone might need, say from marketing or the legal team, to speak about or publish something publicly.
- Manage an external blog, perhaps on *Medium*, and curate and edit posts so they're consistently great.

In a Nutshell

It's notable that DesignOps professionals pay attention to this battery of work sooner and more intently than most ResearchOps professionals might, and the ResearchOps professionals who do give it time often have a background in DesignOps. I've learned a lot about this work by watching DesignOps compatriots from the sidelines. While they made it the centerpiece of their work, in the land of ResearchOps, it was barely touched. But there's good reason for this. As the saying goes, the squeaky wheel gets the grease, and in the world of research operations, participant recruitment, tooling, respect in research, knowledge management, and the support to maintain it all, are *very* squeaky wheels.

continues

In a Nutshell (continued)

Excuses aside, there's discipline to be had in making sure this vital (and quieter) element gets the attention it needs and deserves. Sooner or later, the need to manage the supply and productivity of the central unit of delivery—researchers—will emerge as a critical requirement, and even a hindrance, in scaling research. So, whatever your scale, pay this element well-prioritized attention earlier than you might think. These are the things you should focus on:

- **Partner with people partners and, where possible, integrate with their protocols.** HR professionals have specialized skills and standards, so make sure to partner with them wherever possible.

- **Devise research and ops competency frameworks to use throughout the people management process.** Hiring, onboarding, performing, promoting, growing, and offboarding all have unique operational needs and provide unique opportunities.

- **Put a prioritized plan in place to attract the right talent.** A few efforts done well are better than nothing at all—or trying to do everything. That doesn't work either.

- **Make sure that the people management process is diverse, equitable, and inclusive.** Work with your friendly DEI team, assuming that your organization has one, to do this important and sensitive work. If not, lean on HR or go it alone (and proactively seek out feedback).

- **Put energy into making onboarding a thoughtful experience.** More than just a checklist of to-dos, onboarding should build context and a network for the "onboardee." New-hire onboarding should also integrate seamlessly with your support services.

- **Finally, boost job satisfaction and growth by helping research and ops team members to grow their skills, and their love of the craft.** Provide things for them like easy-to-digest guest talks, courses, book clubs, and speaking opportunities. The ceiling is your creativity (and budget, let's be real!)

CODA: TIME TO THINK

One of the rarest assets of our era is *time to think*. These days, and without knowing much about you, I can make a fair guess that you have an endless stream of messages, emails, meetings, and to-dos to attend to right now. Your heart rate may even have quickened at that thought. In a world that's largely driven by communications in one form or another, it can seem impossible to sequester the time that you need to do *deep work*.[1]

But to deliver scalable research operating systems, you'll need to regularly step back to see the big picture—to shape or review the strategy and the operating system in its entirety—and then you'll need to dive back into the detail to bring it all to functional life. This is true, no matter how big your team or budget, how simple your systems, or how sufficient your support is.

In this book, I've aimed to give you all of the information you need to scale research, but there's one thing that I cannot give you: *time to think*. That is something you must create for yourself. When I worked at Atlassian for a manager who was open to my productivity hacks, I blocked out a week every quarter as a "big picture" week and used it to focus on systems and strategy. No meetings, no messages, no distractions. It was gold. My hope is that you'll also make time to think—it's a discipline, more than anything else.

---

1   The term *deep work* was first coined by bestselling author Cal Newport in his book *Deep Work: Rules for Focused Success in a Distracted World*. Deep work is the ability to focus without distraction on a demanding task.

INDEX

full-service support, 172–173

full-service vs. self-service operating strategies, 57–58, 59–61

Futurama TV series, 176

G

Garfield, Stan, 137

Gawande, Atul, 163

Geison, Chris, 27

General Data Protection Regulation (GDPR), 129, 210, 211

general ledger, 39

Gettier, Edmund, 136

Gibbons, Sarah, 268

Gollan, Casey, 189

Good Strategy Bad Strategy (Rumelt), 28

Gore, Archis, 4, 7

Government Digital Service (GDS), in UK, 56, 257, 280

graduate programs, as talent source, 281–282

gravitational constant, and gravity, 70

group messaging apps, 165–167

guerilla research, 109

H

H1 and H2, 237

halo effect, 284

handbooks, in support services, 161

"The Hare and the Tortoise" (Aesop), 287

Harvard Business Review

10 Must Reads on Strategy, 34

"The Coming of the New Organization" (Drucker), 126

"Communities of Practice: The Organizational Frontier," 143

onboarding report, 285

help desks in support services, 165, 262

HelpDesk, 262

HIPAA (Health Insurance Portability and Accountability Act, 1996), 225

H&M fashion retailer, 46–47, 48, 66

how-to guides, in support services, 161, 162

"Human-Centered Systems Thinking" (IDEO), 50

human resources business partners (HRBP), 275

human resources management (HRM), 274–275

I

IBM Research, Watson, 199

IDEO

collaboration with H&M, 46, 66

"Human-Centered Systems Thinking," 50

Little Book of Design Research Ethics, 213–214, 280

subject-matter experts for ethics, 213

Iger, Bob, 29

implicit knowledge, 141

in-house panels for participant recruitment, 107, 110

in-house research teams and labs, 25–26, 56

incentives for participants, 62, 65–66, 120–121

inclusion, defined in DEI, 283

inclusive recruitment, 123

program management

  compared with project management and operations management, 270–271

  as element of ResearchOps, 73, 75, 77, 90, 194–195, 198

project management, compared with program management and operations management, 270–271

proof of concept (PoC) for new tool, 192

public relations, in building a team brand, 279–280, 282

public relations expert, 258, 281

published support in support services, 161–164

pull system, 62

purchase order (PO), 237

push operating model, 46

push system, 62

push vs. pull operating strategies, 62

PWDR (people who do research), support for, 152, 154

Q

Q1, Q2, Q3, Q4, 237

question log for tooling, 193

R

real-time systems, 4

recruitment. *See* participant recruitment

recruitment brief, 113

red herring, 115

references, as documentation in support services, 162

referrals, for attracting talent, 280

Reichelt, Leisa, 56, 257, 280

repository, in knowledge management, 130

request-and-response service, 63, 173

request forms for research, 261–263

research

  measuring the impact of, 35–37

  need for its own strategy, 30–32

  use of word, 2–3

research backlog, 268–269

research boom, 25–27

research competency framework, 276–278

research culture, 20

research logistics, as decision trees, 15–17

research management framework, 277

research metropolis, building, 11–13

research operating systems

  checklist for scalability, 18–19

  compared with operating system and civic design, 11–13

  culture of, and how to deliver, 20–21

research request forms, 261–263

research strategy, 23–44

  as blind spot among researchers, 25

  characteristics of good strategy, 28–29

  defining, factors in, 31–32

  defining, steps in, 32–33

  need for ResearchOps strategy, 34–35

  operational. *See* operating strategies

  vs. research operations strategy, 22

  vs. strategic research, 24

  strategy as noun and verb, 29–30

in participant recruitment, 111, 119, 122

privacy laws, 225

use of API, 190

right to erasure, 225

The Right Way to Select Technology (Byrne and Gingras), 191

risk management in data privacy, 227

RKM (research knowledge management) Model, 129, 137, 142–146

communities of practice, 142, 143, 144

knowledge services, 142, 145, 146

Rubin, Rick, 67

Rumelt, Richard, 28, 29, 34

runbook, 59, 88

Rybbon, 121

S

Salesforce

recruitment message, 219–220

Service Cloud, 165

scalable systems, characteristics of, 7

scale, as business buzzword, 3–7

defined, 3

efficiency of, 5–6

proportional, not large numbers, 4–5

value of, 6–7

Scaling People: Tactics for Management and Company Building (Johnson), 11

screening, in participant recruitment, 114–115

SECI model, 138–139, 143, 144–145

secondary research, 146

security, 197, 224, 226

Segal, Noam, 38

self-service model for prioritization workflow, 265

self-service support, 172–173

self-service vs. full-service operating strategies, 57–58, 59–60, 96–97

sensitive personal information, 228

service design, 50, 185

Service Design Tools, 172

service desks, 165

The Service Innovation Handbook: Action-Oriented Creative Thinking Toolkit for Service Organizations (Kimbell), 50

service-level agreement (SLA), 63, 165

Shark Tank (TV show), 235

single sign-on (SSO), 190, 223

Situated Learning: Legitimate Peripheral Participation (Lave and Wenger), 143

skills. *See also* research talent

in research competency framework, 276–278

Slack, 165

SMART goals, 28, 246

Snakes and Ladders, 152–153

social recruiting for participants, 109–110

socialization, in SECI model, 139, 144

software-as-a-service (SaaS), 190

solar system, and gravity, 70

source of truth, in support systems, 164, 174

special category data, 228

specialize and optimize, as phase of operational maturity, 84–86

metrics and, 247

planning for tools, 195, 201

in ResearchOps Planning Matrix, 91

T

The Tacit Dimension (Polanyi), 138

tacit knowledge, 138, 141, 142, 157

tags, in knowledge management, 131

Takeuchi, Hirotaka, 138–139, 142

talent. *See* research talent

talent acquisition (TA) in HR, 275, 281

talent events, to attract talent, 281

talent management (TM) in HR, 275

Tango, 121

targeting, in participant
    recruitment, 114

taxonomies
    for data governance, 221
    defined, 130–131
    and failures in knowledge manage-
        ment, 133
    participant, 99

team brand, 279–280

terms of service agreements, 209

text-expander tool, 166

thank you gifts, for participants, 62,
    65–66, 120–121, 122

thank you note, ethics tip for, 218

three phases of operational maturity.
    See operational maturity, three
    phases of

time to think, 291

timeboxed tactics, in support services,
    159, 167

tooling, 177–202
    artificial intelligence (AI), 198–199
    creating a research culture, 179–180
    data interoperability, 189–190

data management terms, 190

planning across three phases of
    maturity, 200–201

planning and steps to onboard a tool,
    190–192

procurement, 181

question log, 193

research tooling blueprint, 184–189

research tooling strategy, 183

ResearchOps, eight elements for
    planning, 196–198

ResearchOps planning matrix,
    193–195

standard tooling criteria, 193

tool maintenance, 181–182

for tracking work over time, 269

types of tools, 180–181

toolkits, in support services, 161

tools and vendors
    as element of ResearchOps, 73, 75, 76,
        90, 194–195, 196, 197

Top Tasks study, 163, 170

Torres, Teresa, 26

tracking research studies, 268–269

training in research. *See* research
    training

transit systems, 13–15, 18–19

Trello, 180, 193

Tremendous, 121

trust
    building a foundation of, 42–44
    in knowledge management, 147

The Tube (London Underground),
    13–15, 18, 19

ACKNOWLEDGMENTS

It takes a small village to produce a book, and there are countless people who have offered their time, wisdom, and support to make *Research That Scales* a reality.

My infinitely patient and smart husband, Glenn Familton, without whom I might have become a skeleton in a task chair. He's been a personal chef, sounding board, coach, and cheerleader, and has rarely flinched when I've asked (for the hundredth time), "Could I read this to you?"

Thank you, Lou Rosenfeld, for thinking that I could write a book, and Marta Justak, for patiently turning me into a better writer and being a friend. Special thanks to my father, John Towsey—everything I know about good business management is down to you—and Debbie Towsey for jumping in when I needed an extra set of eyes. To my dear friend, Elizabeth McGuane, sharing the journey of first-time authorship with you was a joy. We did it. Mom, you've taught me what resilience is all about. And thanks to the Familton clan for cheering me on.

Thank you, Janelle Ward for several inspiring chats and being as courageous as you are, and Leisa Reichelt for providing me with wide-open playgrounds in which to practice my craft.

Smart people reviewed a draft of this book and offered invaluable advice: Behzod Sirjani, Chris Geison, Tim Toy, Alison Jones, Azita Shokrpour, Nick Polochacz, Jo Taylor, Sarit Geertjes, and Penny Rance.[1] And an endless number of people responded to LinkedIn posts, all of which helped me shape the contents of this book—Ian Franklin, your name was reliably there.

Elizabeth Churchill, thank you for writing the foreword, and Meredith Black, Tamara Hale, Nishita Gill, Elizabeth McGuane, Noel Lamb, Matt Gallivan, Ned Dwyer, Tim Toy, and Behzod Sirjani, for offering testimonials.

---

1   Penny Rance publishes an excellent newsletter called "The UX Life Chose Me," which you should subscribe to: https://peneloperance.co.uk/newsletter/

As the founder of Cha Cha Club, a members' club dedicated to ResearchOps professionals, I'm constantly surrounded by *the* people who are shaping the field. "Cha Chas" you keep my finger on the pulse and help me to continually evolve.

Finally, I wrote *Research That Scales* on a land that is not mine, but which has become home—Sydney, Australia. I acknowledge the Gadigal people of the Eora nation, who are the traditional custodians of the land on which this book came to life. I would also like to pay my respects to all Elders past and present, and to the children of today who are the Elders of the future.

 Rosenfeld®

Dear Reader,

Thanks very much for purchasing this book. There's a story behind it and every product we create at Rosenfeld Media.

Since the early 1990s, I've been a User Experience consultant, conference presenter, workshop instructor, and author. (I'm probably best-known for having cowritten *Information Architecture for the Web and Beyond*.) In each of these roles, I've been frustrated by the missed opportunities to apply UX principles and practices.

I started Rosenfeld Media in 2005 with the goal of publishing books whose design and development showed that a publisher could practice what it preached. Since then, we've expanded into producing industry-leading conferences and workshops. In all cases, UX has helped us create better, more successful products—just as you would expect. From employing user research to drive the design of our books and conference programs, to working closely with our conference speakers on their talks, to caring deeply about customer service, we practice what we preach every day.

Please visit rosenfeldmedia.com to learn more about our **conferences**, **workshops**, **free communities**, and **other great resources** that we've made for you. And send your ideas, suggestions, and concerns my way: louis@rosenfeldmedia.com

I'd love to hear from you, and I hope you enjoy the book!

Lou Rosenfeld,
Publisher

RECENT TITLES FROM ROSENFELD MEDIA

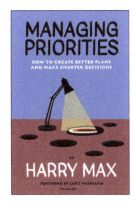

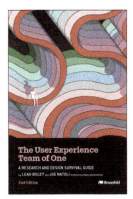

Get a great discount on a Rosenfeld Media book:
visit **rfld.me/deal** to learn more.

SELECTED TITLES FROM ROSENFELD MEDIA

View our full catalog at **rosenfeldmedia.com/books**

ABOUT THE AUTHOR

 **Kate Towsey** is an independent strategist, coach, and advisor. For more than a decade, she has been at the forefront of ResearchOps. The inventor of the widely used acronym PWDR (pau·duh) for "people who do research," Kate also wrote some of the earliest blog posts about setting up research operations, founded the ResearchOps Community, and launched the #WhatisResearchOps movement, putting ResearchOps on the map.

In 2019, she stepped back from the ResearchOps Community and started the Cha Cha Club, a members' club for full-time ResearchOps professionals, which she runs to this day. Over twenty years, Kate has delivered operations in ecommerce, education, government, retail, and tech. She built and managed the Atlassian ResearchOps team, widely recognized as the benchmark for scaling research services and systems to hundreds of people. Learn more at katetowsey.com, or find her on LinkedIn, Medium, or Substack.

www.ingramcontent.com/pod-product-compliance
Lightning Source LLC
La Vergne TN
LVHW011803070326
832902LV00026B/4616